COUNTING

TO

GOD

To discuss this book with the author and other readers
visit the Science, Religion, and Philosophy section of
http://www.attitudemedia.com

COUNTING

TO

GOD

A PERSONAL JOURNEY THROUGH SCIENCE TO BELIEF

Douglas Ell

First Edition

Published in the United States by Attitude Media

2 4 6 8 1 0 9 7 5 3 1

Printed in the United States of America

ISBN 978-0-9632701-8-4

CONTENTS

PART 1

Setting the Stage

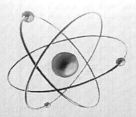

CHAPTER 1

The Great Question

Let me ask you a question. Do you think we live in a meaningless universe, and human beings were created by accident? Or do you think we live in a universe designed and created by a great intelligence, and human beings were designed? Accident or design—that is the question. What do you think?

I think it's a "great question." I think it's *the* great question. It is about the existence of God.

In this book, I'll share new clues. They come not from traditional religion or some televangelist, not from a new prophet or miracle, and certainly not from the government or our mass media. They come from an unexpected source, perhaps the last place you might expect, a place you might think is far from questions of faith. They come from science.

Modern science strongly supports belief in God. That's the message of this book. Contrary to what you may have read, and contrary to what you may believe, modern science strongly supports belief in the God of the Bible. Did you know something caused our universe to come into existence 14 billion years ago? That the laws of physics are fine-tuned to allow the existence of life? That scientists don't have a "mildly plausible" theory for the origin of life by pure chance? That the technology of life is dazzling and more sophisticated than anything human beings have ever created? That the origin of wholly new species remains a mystery? That our Earth is special, and possibly unique, in our entire galaxy in its ability to sustain life over billions of years?

As our scientific knowledge grows, so does the evidence for a designed universe. Each year brings new scientific evidence of wonder. The arrow of scientific discovery points directly to God.

That's not the message of our mass culture, but it is the message of hundreds of scientists, and their numbers are growing. It's a message our culture needs to hear. "When we consider what religion is for mankind, and what science is, it is no exaggeration to say that the future course of history depends upon . . . the relations between them."[1]

You may not be ready for this message. I wasn't. Like millions of others, I thought our modern world had no place for God. I love science; I double majored in math and physics at MIT. Science taught me, or what I mistook for science taught me, that religion is obsolete. I thought only science held the truth, and I thought I had to choose between science and religion.

I wrestled with what seemed to be a conflict. Is modern science consistent with belief in God? After thirty years of thinking, after thirty years of studying philosophy, particle physics, cosmology, evolution, molecular biology, planetary formation, quantum physics, and more, I've learned that modern science is consistent with the Bible, with the three faiths of Abraham—Judaism, Christianity, and Islam. As strange as it may sound, science and mathematics are now the foundation of my faith.

You have a right to believe. Not just a legal right, but an intellectual right. Parts of our society, certain vocal Atheists and much of our mass culture, suggest and sometimes even loudly proclaim that belief in God is outdated and somehow intellectually inferior. That worldview is false. There is a cultural war over your freedom to believe.[2] As one renowned Chinese paleontologist (fossil specialist) said, when asked whether he was concerned that the fossil evidence he had uncovered did not agree with the prevailing theory of evolution: "In China, we can criticize Darwin, but not the government. In America, you can criticize the government, but not Darwin."* That is sad. We pride ourselves on being the land of

---

* This statement was made by J. Y. Chen, during a lecture at the University of Washington. Professor Chen is one of the discoverers of the Maotianshan Shale fossils in southern China, which is generally thought to contain the most exquisitely preserved Cambrian-era fossils in the world. He had described the fossil evidence documenting the sudden emergence of animal life in what is commonly called the "Cambrian explosion." This evidence is discussed in chapter 12. See Steven C.

the free, but in this country scientists are not free to criticize the beliefs of mainstream academia.

Just as you have a right to believe, you have a right not to believe. I'm not out to mock or demean Atheists and Agnostics (but I will poke fun at some of their statements). Many are careful thinkers, very moral and very responsible. I'll present the evidence, their arguments, and my arguments, and encourage you to decide for yourself.

We all have doubts and beliefs. When it comes to God, you have a choice. Atheism is a system of belief, which is why it and the word *Atheist* are capitalized throughout this book. Atheism may not be as detailed as Judaism, Christianity, or Islam, but it can have an equally profound impact on how you live your life. We will examine beliefs about God using reason. I have spent much of my life trying to resolve reason and belief. If you stick with me, I hope to leave you with a deeper, richer appreciation both of our wonderful universe and of God. We will get there with reason and proven scientific facts.

You cannot doubt the existence of God unless you have some faith in the belief that God does not exist. You may say you don't care about God, or have no need for God. If so, you are betting your life that no God exists that could hold you accountable or provide meaning and hope in your life. As Timothy Keller says, "That may or may not be true, but . . . it is quite a leap of faith."[3] Before you take that leap, before you base your life on the belief that no God exists that could make your life meaningful, I urge you to consider the evidence.

I welcome your doubts, and I want to challenge them with science. I want to go right to the core of the new scientific evidence of design in the universe, and thus of the existence of God. To me, it is the most exciting issue of our age. Science is now so advanced that it sheds light, actually a great deal of light, on this ancient and profound question. Let's pose the existence of God as a hypothesis and test it with all the tools of science, all of the experiments, observations, and logic developed over the centuries. Should we believe in God, or should we believe there is

---

Meyer, *Darwin's Doubt: The Explosive Origin of Animal Life and the Case for Intelligent Design* (New York: HarperOne 2013). See also Phillip Johnson, "The Church of Darwin," *Wall Street Journal,* August 16, 1999, also available at http://www.arn.org/docs/johnson/chofdarwin.htm.

no God? We are going to look at the evidence, at the facts, at what I call the "science of belief."

It's a marvelous quest, but it is also one that many are reluctant to embrace. Many theologians shy away, perhaps in part because of a fear that their faith will be damaged by a negative answer and perhaps in part because they think the world of science and the world of faith do not intersect. Many educated persons shun this quest, perhaps in part because they have wrapped themselves in a worldview where the existence of God is literally unthinkable. Many others reject it because, for them, the existence of God would be an inconvenient truth. They do not wish to be held accountable.

For those who do undertake the great quest, there are barriers. First, the current literature, particularly the scientific literature, is often poorly written and ridiculously opaque. Academics write mostly to impress other academics, not for the public. Second, even where the text is readable, it tends to be one-sided, so much so that it may confuse fact and fiction, knowledge and belief, and not acknowledge, or properly state, opposing facts or views. Third, and most insidious, are the barriers deliberately constructed by those who fashion themselves to be our ruling intellectual elite. They have fired, demoted, ostracized, and attacked dozens, perhaps hundreds, of scientists who dare to point out the overwhelming evidence of design, both in physics and in life itself.

This book will help you over these barriers. Let's put the basic facts and arguments "on the table," so to speak. Let's look at the evidence. We are not going to ignore or demean science—to the contrary, we are going to revel in science. I have spent years explaining technical concepts in reasonably plain English, and that is my goal here. I know firsthand that today's scientists are generally bright, educated, and well intentioned. I seek not to disparage the scientific community but to bring you the evidence and promote free debate, and to introduce important scientists whose voices have been lost or ignored by our culture.

I believe in God. Many bright, thoughtful people do not. So let's look at the evidence. As I explain later in this book, it is often very difficult for the existing mainstream paradigm to change as new evidence appears. Mainstream academic economists continued to teach, and preach, the virtues of collective economies even as the Soviet Union imploded and China embraced markets. Sigmund Freud was popular on college campuses long after working psychiatrists had moved past him. It took a long time for behavioral finance to supplant efficient markets theory, despite overwhelming evidence for the former. And so it is with the great question—there is stunning new evidence of design. I will use reason and modern science, not blind faith, to make the case for God. You decide.

CHAPTER 2

The Good News

How can science reveal the existence of God?

It would be an illusion to think that what we are aware of at present is any more than a fraction of the full extent of biological design. In practically every field of fundamental biological research ever-increasing levels of design and complexity are being revealed at an ever-accelerating rate.

Biochemist and noted author Michael Denton,
Evolution: A Theory in Crisis, p. 342

Our LIVES ARE defined by an important choice. A choice we may not even know exists. A choice between wonder and acceptance, between hope and despair, between intellectual freedom and conformity.

Many base their choice on a myth. The myth is that science has somehow displaced religion, that science has somehow triumphed over religion, that science has somehow made religion obsolete. This myth has grown for 150 years. It is a dangerous myth and, to me, the opposite of the truth.

Modern science has revealed a universe of absolute wonder. Wonder in the sense of awe, astonishment, surprise, and admiration. Wonder in what caused our universe to come into being; wonder in why our universe is designed just right for life, wonder in how the incredible complexity of even the simplest life could possibly have arisen. Each year brings new scientific evidence of wonder, facts for which there are essentially no explanations without God, no believable way around the wonder. Contrary to what you may have read, and contrary to what you may believe, science and religion are converging on wonder. The universe is a marvel to behold, and both scientists and religious believers are in awe of its magnificent design.

This book is about a largely unnoticed consensus between the mystic and the scientist. It is about both asking us to look in the same direction, toward a glimpse of a greater reality. It is about wondrous connections among the concepts of number, universe, and God. By observing the universe, through number, we detect evidence of the existence of God.

The riddle of existence is as old as the human race. Why does the universe exist? Is what we see and detect all there is, or is there some type of greater reality, of greater truth? Why do we exist? Can we believe we were put here for a purpose, and if so, what is it?

These are "great questions." There are many questions in our lives, many uncertainties, many doubts. The great questions are in a class by themselves, deeper than all others. Like shadows in the deep, the great questions wait beneath the surface of our lives. When things are well in our lives, when the waters are smooth, it is easy to forget the great questions. But when the waters are rough and the waves threaten to overcome our little boats, the great questions often rise to the surface of our thoughts. We may not say them out loud; we may not even phrase the words. We may be in pain, in danger, alone, or simply confused, and just ask "Why?" or "What now?" When we are in trouble, under stress, when events shake us out of our complacent lives, we have a heightened awareness of the riddle of existence. We are more likely to ask the great questions. We are more likely to step out of our daily patterns and more likely to ask why.

Although ancient, the great questions are more relevant, and more important, today than ever before. They are also deeply personal. How you live your life could depend, perhaps to a great extent, on your personal answers to the great questions. Some devote their lives to a calling they believe comes from God; others mock believers and follow no moral code. Still others invent their own moral code but doubt divine intervention or design.

The good news of this book—the good news of the third millennium—is that modern science strongly supports both belief in a greater reality and belief that both our universe and life itself were designed. The Greek word for "good news" is *gospel*. The first four books of the New Testament are called Gospels because they proclaim the good news of Jesus Christ. Two thousand years later, science—observation and experiment guided by reasoning—alters our lives in so many ways. Yet science supports the faith of Abraham. Two thousand years after the Gospels, science proclaims a new message of hope. To me, it's like science is adding a technical footnote to the Bible. The note is that the hypothesis of God has scientific support. There is clear evidence of design.

The discoveries of modern science are the "new" good news of the third millennium. I will give you seven wonders:

- The universe had a beginning. It has not existed forever. Something caused our universe to come into being 14 billion years ago.
- The universe is fine-tuned for life. If many laws, features, and constants of physics were slightly different, life could not exist.
- Life is a miracle. The simplest organisms contain staggering complexity. Even hard-core Atheists admit there is no plausible chance-based explanation for the origin of life. This book will show, almost to a mathematical certainty, that life could not have arisen by accident or chance.
- Life has technology to dazzle. All life has the same operating system. All life stores information in DNA and transfers that information to build biological machines in basically the same way. There is no evidence there ever was a different operating system; it appears life's operating system somehow sprang into existence 3.5 billion years ago.
- The origin of wholly new species remains a mystery. The fossil record yields few forms that could possibly be transitional. Many biological systems appear to be irreducibly complex, such that they could not have been formed by gradual steps.
- Our Earth is special. There may be no planet better suited for life in our galaxy of hundreds of billions of stars.
- Quantum physics challenges traditional concepts of matter, space, and time, and invites a different way of looking at reality. We will consider one such way—fully consistent with verified experimental facts—that strongly points to the existence of God.

Each of these wonders is scientific support for the hypothesis of God. If the God of the Bible exists, then we would expect the following:

- the universe was created;
- the universe was designed for life;
- the creation of life was a miracle; and
- life has amazing technology.

We could also expect to find that

- new life-forms were not created by accidental events and natural selection alone; and
- our Earth is special.

Finally, we just might find

- scientific evidence that traditional, materialistic views of reality are not complete.

We are each entitled to our own views, our own perceptions, our own reality. I respect Atheists and Agnostics. Although I do not share their point of view, I am not troubled by Atheists; I can see how a person could conclude that there is no God. Although I do not share their point of view, I am not troubled by Agnostics; I can see how a person could conclude that one does not know, and perhaps will never know, whether God exists. Each person is entitled to his or her own point of view, his or her personal answers to the great questions.

I do reject the misuse of science in the debate. In my view, science and mathematics strongly support belief in God. They do not compel belief. But the proposition, advanced by a small but vocal minority of Atheists, that science somehow reveals the folly of religion, is wholly false and dangerous. And I most strongly reject the animosity—even hatred—that has arisen against those who dare to suggest that science supports belief in God. Read this book; understand the facts and basic concepts; decide for yourself.

Some doubt the existence of God because of the seemingly overwhelming problems of our world. Those concerns, those doubts, are generally beyond the scope of this book. Others have addressed them eloquently—for example, Timothy Keller in *The Reason for God*.[1] We're going to focus on the scientific facts supporting belief in God—the science of belief. In the last chapter, when we're "Connecting the Dots," I'll give you my personal thoughts on why the world may be the way it is.

This book follows my thirty-year journey to reconcile science and religion. There's a lot to cover, and I'm going to approach it in a way that, after considerable trial and error, seems best suited to weave the different themes together.

This initial part sets the stage; it introduces the science, the concepts, and the history of the "great debate." I'll tell you about my journey, because your journey may be similar to mine. We'll look at what is science and what is religion, and why they are allies, not enemies, in the search for truth. We'll consider the concept of a paradigm and why many scientists choose to ignore, or seek to explain away, the current evidence of design, and we'll see how key paradigms about space and time have shifted over the course of civilization. We'll look at the history of the great debate over the existence of God and at the latest tools scientists use to detect evidence of design in physics and in life.

Part 2 reveals the science of belief, the existing evidence, and generally follows the order in which I explored each subject in my journey. We'll start with the creation of the universe and the fine-tuning of the laws of physics. We'll note scientific and logical problems in the concept of an infinite multiverse. Next comes the mystery of the origin of life, the technology of life as revealed by molecular biology, and the puzzles of macroevolution. Then we journey back to physics and recent evidence that our Earth is special. This part ends by suggesting a seventh wonder of modern science in the mathematical nature of the universe combined with the nonmaterial facts of quantum physics.

Part 3 sums it up. It begins with a review of the scientific arguments and counterarguments for belief. It ends with my final thoughts, how I connect the dots after thirty years.

Where possible, I'll try to keep it light. We're only talking about the existence of God and the meaning of life, using concepts not much more complicated than general relativity and quantum physics. You do not need to fill out a single government form.

More seriously, I want to reassure you that this book does not require heavy math. There are no equations to solve. The primary concept of number in this book is exponents. We are going to be comparing ridiculously large numbers, like stars in the universe to possible combinations of amino acids, and we need exponents to do that. Appendix A contains a simple, easy-to-understand refresher. Trust me; you can follow the math in this book. If you can count to seven, you can count to God.

I wrote this book to give my children, and hopefully you, a choice. I want to help you look at the evidence with an open mind. Again, I want to go right to the core of whether there is evidence of design in the universe, and thus evidence of a greater reality. I want to take you on a journey through number, universe, and God, and I hope to leave you with a sense of wonder—a sense of awe, astonishment, surprise, and admiration. To me, science and religion are not opposites. Science and religion are two paths on the road to truth, and both point in the direction of wonder.

This book uses the word "number" to include all of mathematics—all of its patterns, equations, symmetries, and beauty. For me "number" is mathematics as an art form. I sometimes associate the word "mathematics" with routine drudgery—as in "go do your math homework." When I think of the art and beauty of mathematics, of stunning patterns and connections, of the triumphs of logical reasoning, I use the word "number." In that sense, this book links number, universe, and God. In that sense, we will use number to detect evidence of God.

This is the story of my journey through science to belief.

CHAPTER 3

A Personal Journey

Why me?

If you study science deep enough and long enough, it will force you to believe in God.

Lord Kelvin

Tʜɪs ᴄʜᴀᴘᴛᴇʀ ɪs about my personal journey, the twisting and the process that I went through, over decades, seeking the truth. It is not required reading; the wonders of modern science are in later chapters. But perhaps your journey is similar to mine, and perhaps the doubts I faced are similar to your doubts.

I cannot say I believe in God solely because of science and mathematics, but science and mathematics are now the foundation of my faith. I had thought the opposite; I thought science and mathematics would guide me away from ancient myths toward some modern, more rational reality. That is not what happened. It is as though, while following road signs to logic, I came to a space of profound mystery.

When I was young, I liked numbers. I really liked numbers. They were magic, and they were my friends. Their precision was comforting, and their relationships are still beautiful. I was so proud when, as a very young boy, I realized I knew all the names for all the integers between one and one million. I remember asking my grandmother if she wanted to hear me count to one million, and she, being incredibly sweet, made the serious mistake of saying yes. So I started counting, and as time went by and even my grandmother's rapt attention began to fade, I realized I needed to speed things up. But that's the kind of kid I was, very much lost in my own thoughts. Another thing I actually thought was fun in early grammar school was to write down two very large numbers—perhaps twenty or more digits each—and lie on the living room floor and multiply them by hand. Yep, quality fun.

When I got a little older, I began to learn about our universe, and I was fascinated. I devoured everything I could find about stars, galaxies, natural history (dinosaurs!), and theories of creation. I read every book in the local library on math and science.

God was harder. I was very troubled, for many years, by the perception that belief in God is incompatible with logic and science. I now reject that perception completely. In this book, I weave number, universe, and God into one reality, a greater reality, my personal reconciliation. After a journey through logic and the latest scientific discoveries, I have arrived at the pillars of the faiths of Abraham. It is a trek through logic to wonder, to awe at the majesty of creation. I invite you to take it with me.

Of all the "great questions," one has preoccupied humanity since the dawn of time. Are we—and all of existence—here by accident or by design? This great question drives philosophy. It overshadows morality and ethics. It anchors Western religion.

For reasons I cannot fully explain, this great question has taken hold of my life. I have wrestled with it for decades, and it has spun me in circles. I do not claim to have conquered it; perhaps, it can never be fully conquered. But I have glimpsed beauty profound; I have sweated drops of astonishment. And I find it odd, genuinely mystifyingly odd, that in all my spinning, I have landed in the very place from which I started, with the same awe and the same sense of childlike wonder with which I began. In my own way, in my own contorted, clumsy, and confused manner, I believe I have wrestled with God.

What is God? What does it mean when someone says they "believe in God" or they "don't believe in God"? I've struggled with that question since I was small. I first thought of God as a kind old man with magic powers. In Sunday school I learned that God used to be cranky—back when the Old Testament books of the Bible were written. My earliest religious philosophy was simple: At one point God got fed up and drowned everybody except for Noah and his family. Lucky for us, Jesus Christ was born, God cheered up, and things have been better ever since.

When I was very young, my church gave me a Bible, and I decided to read it. I didn't get far. Some of the first stories in the Old Testament gave me problems. I wanted to believe, so when I heard about people who lived almost a thousand years each—Methuselah and others with weird names—I asked my mother. I don't remember what she said, but I finally decided it was probably a mistake of some kind. I learned that the Bible was copied by hand and translated from different languages. I was having enormous trouble with handwriting—I got straight Ds in handwriting in elementary school—so maybe the numbers were written wrong or the person who copied them couldn't understand the handwriting.*

More important, it didn't seem to matter how old those guys were when they died. The Bible doesn't say that they did anything special or were important for any reason, except possibly as ancestors. As far as I could tell, they were just really old when they died. So even though I found it hard to believe that people could have lived almost one thousand years, my faith was tested and it held.

Not for long. The next chapter in Genesis tells the story of Noah and the great flood. The basic story, as you probably know, is that people were acting bad, so God got angry and decided to drown the whole world, except for Noah, whom God liked, and Noah's family. God warns Noah and tells him to build this really big boat—called an "ark"—and gather two, a male and a female, of every animal on Earth so that each type of animal—each species—could survive the flood. Noah somehow does all this, and then it

---

* A few years later, I thought of another possible solution to the Methuselah problem. Perhaps they counted months instead of years. I saw that if Methuselah had lived 969 months instead of 969 years (a different type of copying error), then you would have to divide by 12 to get his age in years. (Actually, they might have used lunar months or cycles, which are about 29.5 days, but that doesn't change the answer much.) I saw that 969 divided by 12 would have made him a little more than 80 years old when he died. I had met people in their eighties and certainly could see why someone writing the Bible would think it very special that anyone lived that long. So my "advanced" grammar school theory was that the Methuselah story was based on a misunderstanding. Someone meant that Methuselah lived for 969 months, and in retelling or recopying someone else goofed, and we've been stuck with the 969 years "misprint" ever since.

rains like you know what for 40 days and 40 nights until the whole earth is covered with water and everything else drowns. The Bible says the water covered the highest mountains. Meanwhile, Noah and his family and the animals he collected float merrily along. According to the Bible, the water did not start going down for 150 days!

Right from the beginning, I had problems with this story, and they got worse as I got older. At the zoo I learned there are lots of different animals. For starters, there are dozens and dozens of different kinds of antelopes. (It always was a mystery to me why zoos needed so many kinds of antelopes. To me, you see one antelope, you've seen them all.) And was I supposed to believe the lions and tigers walked calmly onto the boat and didn't bother anyone for 190 days or whatever? Sure.

While struggling to fit all these animals onto one boat, and thinking about how they could all survive, I realized there was a big problem with bugs and some other creatures. There are millions of different kinds of beetles, and way too many different kinds of spiders, flies, and so on. Did Noah collect two of each? How could he tell male from female? Did these millions of bugs just sneak onto the ark and the Bible was too polite to mention it? If the bugs weren't on the ark, how did they keep from drowning? And why would Noah or any other sane person even want to save the bugs? Did God make him do it? Why? What about snakes?

I also couldn't figure out what happened to all the water. If the whole Earth was covered by water, water so deep that it covered the highest mountains, where did it go? The story made it sound like Earth was a bathtub. It took 40 days and 40 nights to fill it up and cover everything. Then God pulls the plug, and the water disappears.

As you can see, I found it hard to believe the story of Noah and his ark. It's a fascinating story of God calling a chosen people, and I realize now it holds meaning for many people. But my young faith "choked" on this story, and I stopped reading the Bible. I tried to take the text literally, and my "copying mistake" idea didn't seem to work. Did God drown the whole world or not? I also didn't understand the point. How many people did God drown, and exactly what did they do to deserve being drowned?

I began to doubt God and the Bible. As I got older and entered high school, I started reading books to break up my daydreaming in the back of the classroom. Many were James Bond or science fiction, but there were others. I read Karl Marx's statement that "religion is the opiate of the masses," by which I think he meant that religion gives ignorant people false hopes that make their suffering bearable. Another quote I remember is "If God didn't exist, man would have to create him." I took this to mean that religion and God were made up by people to explain things for which they had no other answers.

My belief in God got weaker. It didn't help that I was learning more and more about science's explanations of why things are and how they work. Science sure seemed to have a lot of answers, and the answers seemed to make sense, at least more sense than that story of Noah and the ark, which really didn't explain anything, at least not to me.

At one point in high school I wrote my minister and told him I didn't believe in God. I'm not sure exactly why, but he probably had invited me on some youth group trip. He wrote back a nice note and sent me a short book. All I remember of the book is its title—*Your God Is Too Small.* Looking back, I think my minister hit the nail on the head; he knew exactly what my problem was. My image of God as a kindly father figure, as sort of a year-round Santa Claus, was too small. It had to go. The problem was that I wasn't ready to shed that image and move on, perhaps because I wasn't sure what to replace it with. And so my belief faded, so much that only God could have held out hope of recovery.

I became the perpetual student of my family. It seemed clear that getting a job would interfere with my daydreaming. I considered myself an Atheist—a person who doesn't believe in God. I went to college at the Massachusetts Institute of Technology—or MIT, as it is usually called—which was and still is a pretty fancy place with lots of smart people and amazing classes. I spent more than ten years in college and graduate schools. I got an undergraduate degree in math and physics from MIT, and then a graduate degree in theoretical mathematics from the University of Maryland. I obtained a law degree, which fit my need to carefully and logically dissect every issue. It seemed you could explain just about everything with logic and science. It seemed God had no place in our modern world. I treated God like a joke.

They say all good things must end. Despite my best efforts, the day finally came when I had to get a serious job. Life changed pretty fast after that.

I didn't really go back to church until my early thirties, after my son was born. Like many people, I went to occasional services for the holidays, like Christmas and Easter, but my heart—my soul—wasn't in it. I had a nasty attitude on the inside. I shut God out.

A year after my son was born, my wife decided we should join a church and get him baptized. I figured it couldn't hurt. Even though I knew—or I thought I knew—that God didn't exist, there didn't seem to be much downside to a few visits to church. I measured the pros and cons. On the one hand, if there is no God, as I thought, then all I've done is waste a little time to please my wife. No big deal. On the other hand, if there really is a God, I had it covered. I thought I was smart. I thought I had outsmarted God.

No chance. My plan fell apart our first Sunday. One problem was that there were some really nice people at the church we went to. Some of them were younger couples like us, and we started to become friends. Another problem was that the minister was great. To my surprise, I enjoyed his sermons. I was most impressed with his sense of inner peace. I could sense a bit of that same wonderful calmness and peace in some of the other people at church. I didn't know what it was. I just knew I wanted it. Something was missing from my life, and I wanted it.

Looking back, I honestly think God decided to give me, out of pure, undeserved grace, the ability to be aware of and recognize the inner peace I was missing. I am not asking you to "believe" anything about my story or about inner peace; I am just describing what happened to me. Many people would have seen nothing special in the minister or other persons at that church. For some reason, some truly unknown reason, I quickly knew that something incredibly important was missing from my life. I realized then that no amount of professional success or monetary affluence would ever be enough. It would never be fully satisfying. I wanted something more, something much more special and valuable.

We started to go to church regularly. We met wonderful new friends, but I struggled with old doubts. As a somewhat educated person—I sure spent a lot of time in various schools—how did it all fit together? How could God stand up to logic and science? What is God? Where is God? Why is there a universe? Why are we here? What are we supposed to do with our lives? The great questions had me pinned to the mat.

In my midforties, my wife and I decided to take a chance. Though I continued to work as a lawyer in Washington, DC, we decided to move 380 miles away, to Martha's Vineyard, an island just south of Cape Cod in Massachusetts. We were tired of suburban life, and we wanted our kids to grow up in a small town. We picked Edgartown because my mother and stepfather had retired there. At the time, I was traveling quite a bit for business, and I thought it wouldn't be much harder to travel from the Vineyard instead of Washington, DC. This was just before the island became somewhat fashionable, but that's another story. I remember discussing this crazy move—to place my home and my family 380 miles and two airplane flights away from my job—with family, friends, and colleagues. Every person told me I was crazy (even my mother); some told me I could be committing professional suicide. For some reason, I didn't care; I made the move. I remember thinking that if I stayed in the Washington, DC, suburbs I would always be missing something.

The move was a gamble, but it was ultimately worth it. As Robert Frost put it, I took the road less traveled by, and that has made all the difference.

I went from spending a bit of time on planes to spending a lot of time on planes; most weeks, I was on four flights or more. Much of this time I spent reading, especially nonfiction books on mathematics, science, and religion. To me, there were patterns and connections among all of these areas. This book is about those connections, and the answers that I came to during years of struggling. It is about my journey from Atheism to belief.

I am certainly not the first to suggest connections among mathematics, science, and religion. Hundreds of other books have explored various subsets of these issues—most notable may be the clashing books for and against Darwin's theory of evolution (see chapter 12). What I have done is give a great deal of thought as to how it all fits together. You might think of the various facts, theories, and arguments as "dots" in a picture to be revealed. What I hope to do is to connect the dots in a way that may help you see the larger picture, may help you come to your own personal views, may help you find your own answers to the great questions.

It is also my intent to explore these controversial issues in a calm and factual manner. Let's put the facts on the table—the latest evidence, the latest scientific analysis. Many books on these subjects talk past each other. They ignore or berate opposing views (again, many of the books about Darwin's theory of evolution demonstrate this tendency). Yes, I have my views, but I have taken steps to verify the accuracy of every scientific statement in this book. I have attempted to avoid unproductive or gratuitous criticism. I do not wish to mock or demean Atheists and Agnostics. Every person is entitled to his or her own answers to the great questions. I do openly challenge those who deride religion as being contrary to or in conflict with modern science.

This book weaves together technical, religious, and philosophical discussions. I hope that my background, and my independence from traditional academia, make me a good weaver. I hope to convey technical ideas and philosophic concepts in reasonably plain English.

Before we get to the science (part 2), we're going to look at what is science, and what is religion. We'll look at the concept of "paradigm shift," how accepted concepts of space and time and the universe have shifted over human history, and how difficult it is to change the current bias against the existence of design in the universe. We'll review the history of the "great debate" over the existence of God and the latest scientific tools for detecting the existence of God.

CHAPTER 4

Religion versus Scientism

What is religion, and what is science?

The scientist's religious feeling takes the form of a rapturous amazement at the harmony of natural law, which reveals an intelligence of such superiority that, in comparison with it, the highest intelligence of human beings is an utterly insignificant reflection. This feeling is the guiding principle of his life and work.

ALBERT EINSTEIN

If it disagrees with experiment, it's wrong. In that simple statement is the key to science.

RICHARD FEYNMAN, Lectures at Cornell

PEOPLE USE THE WORD "religion" in different ways, and, if you want to get technical about it, it's not always clear when something becomes a religion. But for most people in the United States, and actually for more than half of the people in the world, their "religion" is a tradition of faith that traces back 3,800 years to a man called Abraham.* Abraham

---

* At one time many scholars doubted Abraham was a real person. There were no known places to match the biblical records of the route he took and the battles he fought. Modern archeology has filled in those gaps and more. The biblical towns have been uncovered and even the name of Abraham has been found written in stone. Today, it is commonly accepted that Abraham was a real person.

23

is considered the father of three major world religions—Judaism, Christianity, and Islam. They are collectively referred to in this book as the faiths of Abraham. Today there are perhaps 14 million Jews, 2.3 billion Christians, and 1.6 billion Muslims.[1]

The faiths of Abraham make claims about the universe that many other religions do not make. The faiths of Abraham claim that everything in our reality was created by, was designed by, and is subject to the control of an intelligent being in a different reality.* It is a shocking claim; it only seems less so because civilization has had 3,800 years to get used to the idea. It began as an assertion of pure faith, thousands of years before humanity would even consider the possibility of experimental verification. This intelligence, this being who they claim designed, created, and controls our reality, is the "God" of the faiths of Abraham, the God of the Bible. The Hebrew Bible contains fantastic stories of Abraham speaking with God, and even of his grandson Jacob wrestling with God or an angel of God.[2]

Religion helps people understand their world. But some would argue, indeed many today would strongly argue, that the biblical stories and the faiths of Abraham can no longer be taken seriously. To some, they are an unwelcome caveman relic of a primitive past—like finding a bearskin loincloth in a men's clothing store. A competing and powerful view of reality has arisen. It claims to be based on modern science, but it is not, and that is the heart of this book. It is a new and competing system of belief.

One name given to this new belief system is "Scientism." Another is "Naturalism." Stated most simply, Scientism (aka Naturalism) is the belief that everything can or eventually will be explained by science. The PBS series *Faith and Reason* puts it this way: "In essence, Scientism sees science as the absolute and only justifiable access to the truth." Scientism is the belief that only this reality, and only the natural laws of this reality, can be true. Scientism rejects any possible validity of religious or metaphysical inquiries. There can be no greater truth, no greater reality. Scientism "is the view that the physical world is a self-contained system that works by blind, unbroken natural laws."[3] By definition, Scientism rejects the existence of God.

---

* According to Jewish tradition, Abraham's father, Terach, was an idol merchant. Abraham questioned his father's faith and ultimately came to believe that the universe was created by a single intelligence. God asked Abraham to leave his home and become a nomadic wanderer.

Scientism is not science. Science is the observation, experimental investigation, and explanation of natural phenomena.* Scientism puts a box around science and says you can't look outside the box for truth, even when you ask why the stuff in the box is there or how it came to be. According to Scientism, the box is the box is the box. There is nothing else; there are no truths other than the truths of science.

But that is a *belief.* When engaged in science, when observing the natural world and seeking explanations of phenomena, you do not have to believe that the natural laws you will observe will ultimately explain everything, including the riddle of existence. Of course, almost all of the time, the vast body of knowledge we call science will perfectly and adequately explain what you observe. If one wishes to calculate rocket trajectories or predict chemical reactions, there is generally no need to suggest metaphysical explanations or processes.† We realize no magic words or ceremony will turn base metals into gold. The "box" of modern science is large; it can predict weather, galaxy formation, crop yields, and mutation probabilities, among countless other items. But is there more to reality? Is science the only truth? A few questions, a few great questions, ask what is outside the box. The great questions are beyond science.

Scientism says there can be no greater reality. Scientism is a powerful belief. Perhaps without fully realizing what is at stake, a large part of our "popular" Western culture has swallowed it whole. It has become the prevailing paradigm of existence, and it is bleak. It goes something like this: There is no other reality. There is no God. We are here solely because of random, purposeless events. Our universe was not designed. Our Earth is not special. Life has no purpose. Humanity owes its existence solely to the random, purposeless, accidental creation of a living organism billions of years ago and countless random mutations driven

---

* The first definition of science in the *American Heritage* Dictionary is "[t]he observation, identification, description, experimental investigation, and theoretical explanation of natural phenomena." So explanations of natural phenomena are clearly science, such as Newton's theory of universal gravitation, or Einstein's theory of general relativity, or mathematical evidence of design in the creation of the universe and life itself.

† A famous example of arguably going beyond traditional science was the prediction in 1952–53 by Fred Hoyle of a key quality of carbon atoms. Hoyle suggested this quality—a nuclear energy "resonance"—to explain how carbon was created inside stars. In essence, he made this prediction not on the basis of scientific evidence or theory but rather on the assumption that things had to be just right for carbon, and living creatures, to exist.

by natural selection. We are "just a chemical scum on a moderate-sized planet, orbiting around a very average star in the outer suburb of one among a hundred billion galaxies."[4] There is not, and has never been, a greater purpose or a greater meaning to life. From this view, "The more the universe seems comprehensible, the more it seems pointless."[5]

It is a bleak but powerful philosophy. It has its own high priests and zealots, who will use any means necessary to defend it. In most major universities, a teacher of the sciences who openly challenges this bleak philosophy is not likely to get tenure. Our powerful modern media—newspapers, magazines, television, and movies—continually reinforce this anti-faith worldview. Scientism is a system of belief, but it is not a "faith" in the traditional sense. Scientism is the direct opposite to traditional faith; it is belief in "anti-faith."

The truth, as this book will show, is that modern science contradicts the bleak anti-faith worldview of Scientism. Allegorically speaking, the anti-faith worldview—Scientism (aka Naturalism)—has no clothes. It is NOT consistent with modern science. As philosopher Alvin Plantinga puts it, "[T]here is superficial conflict but deep concord between science and theistic religion, but superficial concord and deep conflict between science and naturalism."[6]

As we begin this third millennium, there is amazing scientific evidence that our reality is not all there is and that both our universe and life itself were designed. I see in this evidence a powerful cause for hope. Religious belief in a created universe has become accepted scientific fact. As many have pointed out, the first words of the book of Genesis describe in general terms the sequence that led to human beings. We will not debate here how many days it took.*

Science now tells us our universe was created and is fine-tuned for life. Science now tells us Earth is a special planet. Science now tells us, essentially to a mathematical certainty, that life could not have arisen by chance and that the design of human beings could not have been a meaningless, random event. I think science is telling us there is a greater reality. I think science suggests that everything in our universe originated as an idea in the mind of God.

---

* The original Hebrew word is "yom," and one of its meanings is an indefinite period of time. See chapter 16.

There is plenty of intellectual room for wonder, and that is wondrous news. If you choose to believe in some kind of a greater reality or in some kind of purpose to the universe, you need not abandon science or logic. You can hold your head high in debate, and science and logic will be your allies. You have been given freedom—intellectual and scientific freedom—to believe or not to believe. You have been given seven pillars of scientific support for the existence of God. You have been given evidence of God. That is very good news.

But for now, at least, you will not find this message of hope reported in the popular media. It will not make the evening news, or tomorrow's *New York Times*. It is contrary to the prevailing worldview, and modern culture is not ready to accept it. Many people will be troubled by this book. They will mock and dismiss it, perhaps without even opening the cover. They will be rigid in their view that religion is contrary to science and will not be open to the evidence, to scientific facts that strongly point in the opposite direction.

To learn, you must begin with an open mind. This book will not ask you to use faith to overcome gaps in logic. It will not ask you to ignore scientific facts; it will ask you to take them seriously.

A lot has been written on what is religion and what is science, most of which is of little help here. This book focuses on the beliefs of the faiths of Abraham, to the extent they can be stated as facts subject to experiment—such as beliefs about creation, the universe, Earth, and life. As for science, I like the way physicist Richard Feynman put it: "If it disagrees with experiment, it's wrong. In that simple statement is the key to science."[7] As we will see, these fundamental Abrahamic beliefs agree with experiment and observation, with true science. The experimental facts disagree with the belief system of Scientism.

I am in awe of the discoveries of modern science, and with the scientific method for investigating our world. I believe we should apply human observation and logic to the fullest when seeking to understand the universe. I commend the multitude of scientists who, over the centuries, have made such a profound difference in our lives.

The time has come to examine the common ground between science and religion with reason, and not emotion. Science and religion are different ways to the truth and approach the great questions by different paths. We must move beyond not only the ignorant distrust of science but also the modern condemnation by our popular media and certain groups of religious beliefs. Persons on each side despise the worldview of the other. Some religious traditionalists treat science as heresy. Some scientists mock religious beliefs. Many on both sides refuse to acknowledge the perhaps shocking consensus that is beginning to emerge. Many theologians are not comfortable with efforts to detect scientific evidence of God in the universe. Many scientists mistakenly reject such evidence as unscientific, because it threatens their system of belief, their Scientism.

It is unthinkable, perhaps horrifying, to many scientists and religious persons that the other discipline is headed in the same direction. Yet that is exactly what I see emerging. Like it or not, science and religion have become allies in the search for ultimate truth. They have become interconnected paths to understand the universe. Science and religion are converging on wonder.

There are new connections among the ancient concepts of number, universe, and God. Knowing these powerful and unexpected connections could make a difference in your life. Though the belief system of Scientism, belief in anti-faith, has become the prevailing paradigm—the accepted worldview—of our modern culture and is strongly advocated by a vocal minority of Atheists, Scientism is not accepted by most people. In the United States, a strong majority of adults believe in God. In a June 2011 Gallup poll, 92 percent of adults answered "yes" when asked, "Do you believe in God."[8] Belief was less—but always above 80 percent— among young people and liberals.

Despite this, many religious believers, particularly well-educated believers, are painfully aware that they are not in sync with the popular paradigm. They often feel intellectually oppressed by modern culture, uneasy about their beliefs, and have a misplaced perception that those beliefs are contrary to science. In a sense, they feel persecuted. Some followers of the faiths of Abraham think they must reject modern science. As I will show, that is not true—belief and science are now allies.

I wrote this book so you may have the intellectual freedom to believe. Faith is a gift. If your mind is closed, no book can give it to you. If open, this book may help you understand that science and reason do not require you to abandon faith. This book may give you a glimmer of hope and an intellectual foundation for belief. There truly is a strong scientific basis for wonder—for awe, astonishment, surprise, and admiration at the miracle of existence.

This scientific basis for wonder is increasing. Not so long ago, people thought our universe had always existed; now we know our universe was created, and that space and time are connected and flexible. Not so long ago, people thought our bodies were made up largely of some simple, jellylike substance; now we know our bodies have trillions of specialized and interconnected cells, each of which is like a complex, three-dimensional city with libraries, factories, trucks, and highways. More recently, we learned a mysterious force is pushing galaxies away from us, and we learned DNA contains multiple levels of information. Modern science continues to reveal the wonder of existence.

You have a choice. You can accept the dogma of Scientism as fact and believe the universe is an accident, without meaning and without purpose, and live your life that way. Or you can use the gift of reason to consider new evidence, evidence that just might lead you to believe in a designed universe of absolute wonder and evidence that just might let you live your life with meaning, with purpose, and with a sense of a greater reality, in awe of life's mysteries and designs. Choose well: it's your life.

CHAPTER 5

Paradigm Blindness

If it's true, why don't people see it?

A man hears what he wants to hear, and disregards the rest.

PAUL SIMON, from "The Boxer"

You MIGHT THINK, or perhaps you would like to think, that scientists are always open to and always seeking out new ways of looking at and understanding our world. At the level of specific individual facts and narrow, specialized areas and theories, that is generally true. But scientists are people, and like all people, they tend to approach problems and issues using the techniques and assumptions they have learned. They strive to fit scientific facts into the concepts of reality they have learned and adopted. They too suffer from "confirmation bias"—a tendency to favor information that confirms their internal beliefs and assumptions. Confirmation bias is like an internal "yes man."

In his 1962 book, *The Structure of Scientific Revolutions,* Thomas Kuhn described how difficult it is to challenge or change broad scientific concepts. His book was named by the *New York Times* in 1987 as one of the 100 most influential books since the Second World War. It has been cited in over 28,000 scientific and scholarly articles.

Kuhn introduced the term "paradigm shift." He observed that scientific development generally does not take place in a smooth, linear manner. Instead, scientists create worldviews, or "paradigms," to solve problems. A paradigm is a way of looking at the world and approaching problems. Within the paradigm, scientists develop ways for addressing and solving problems, and it becomes hard for them to approach problems any other way. Kuhn wrote: "In a sense that I am unable to explicate further, the proponents of competing paradigms practice their trades in different worlds."

This is true today. Some scientists work within a random, pointless universe, while others work within a universe of design and wonder. A vast gulf separates the two views. The concept of paradigm shift helps to explain why so many intelligent people are culturally blinded to the religious implications of modern science.

Philosopher Arthur Schopenhauer is said to have written: "All truth passes through three stages. First, it is ridiculed. Second, it is violently opposed. Third, it is accepted as self-evident." Evidence of design in the universe is at the second stage. It faces violent, virulent opposition.

Some of the greatest paradigm shifts in history have involved concepts of space and time. The examples below illustrate the concept of a worldview/paradigm and how difficult it is to challenge or change an existing paradigm, particularly for well-respected scientists. These scientists have been schooled in the paradigm; they have secured tenure within it and have wrapped themselves in it both professionally and personally. They have a huge investment in the existing paradigm; to a large extent, they have defined themselves by it, and they cannot see beyond it.

The examples below introduce some of the science we will study in part 2. They also illustrate a theme of chapter 14—concepts of number have not only changed our understanding of the universe but also have become how we understand the universe.

Aristotle

Aristotle (384 BCE to 322 BCE) is a giant in the history of civilization. According to *The Encyclopedia Britannica*, "Aristotle, more than any other thinker, determined the orientation and the content of Western intellectual history." Aristotle explored the natural world largely through logic, which he believed to be supreme above all. Aristotle wrote treatises on many subjects, and he is said to have founded logic and biology. His views remained influential for two thousand years, and in an important way Aristotle helped to create Western civilization.* Aristotle believed in a god of sorts, but not a moral or compassionate god and certainly not the God of the Bible.

Aristotle believed the universe was infinite and eternal—limitless in size and had always existed—with Earth as its center. Other heavenly bodies—the Sun, the Moon, the stars, and the planets—were all thought to revolve around Earth. They were thought to be divine and capable of moving only in perfect circles. Elaborate models were created with Earth at the center and various crystal shells rotating around Earth. One of the shells was the fixed stars. Other shells accounted for the planets.

One problem was that some heavenly bodies did not fit well within the shell model. The planets seemed to go back and forth relative to the stars, in strange paths that had little if any relation to perfect circles with Earth at the center. For example, at certain times the planet Mars traces a loop in the sky against the "fixed" background stars.

To accommodate such strange motions, this ancient model—known as the Ptolemaic system—was repeatedly made more complex with circles on top of circles. The resulting chaos was less than divine. Some believed that "no system so cumbersome and inaccurate as the Ptolemaic could possibly be true of nature."† Despite these problems, the Ptolemaic system continued as the reigning paradigm for more than two thousand years.

Copernicus

After 1,900 years, Nicolaus Copernicus (1473–1543) challenged the classical Greek paradigm and the Ptolemaic system. Copernicus was educated in Italy but lived primarily in what is now Poland. He dabbled in many fields—he was a mathematician, a military leader, a diplomat, a governor, and even a canon in the Catholic Church‡—but

---

* Aristotle was also the tutor of Alexander the Great.

† Kuhn attributes this statement to Domenico da Novara, a colleague of Copernicus. Kuhn, *The Structure of Scientific Revolutions*, pbk ed. (Chicago: University of Chicago Press, 1970), p. 69.

‡ Owen Gingerich of Harvard, who has written a book on Copernicus, tells me that Copernicus was in charge of one of the altars at the Archcathedral Basilica in Frombork, Poland.

he is almost always remembered for his work in astronomy and his challenge to Aristotle.

Based on astronomical observations, Copernicus suggested the Sun was the center of the cosmos, and Earth and the other planets revolved around the Sun. It may be the most shocking, revolutionary, theory in the history of science.* At once it provided a radically new way to look at a variety of facts—the varying brightness of the planets (now explained in part by differences in distance from Earth), the closeness of Venus and Mercury to the Sun (now explained by smaller orbits), the varying speeds of the planets (Copernicus was able to calculate the order of the planets from their relative speeds), and so on.

While elegant in many respects, the new Copernican model had its own serious problems. How could it be, if Earth revolved around the Sun, that the stars remained fixed? Was it really possible that the stars were almost unthinkably far away and Earth was adrift in immense space? And if the motion of the stars was caused by Earth rotating, why didn't everything immediately fly off the surface of Earth? There was no known principle of physics that could explain the stationary nature of objects on Earth.

Copernicus delayed publishing for several reasons, one of which was to avoid the scorn "to which he would expose himself on account of the novelty and incomprehensibility of his theses."[1] Copernicus's theory "has become a symbol of defiance against the Church's teachings. But it was the Church itself that had invited Copernicus to come up with a new math for the motions in the heavens."[2] Sometime before 1514 Copernicus circulated a forty-page summary to friends, who spread the word. Pope Clement VII and various Catholic cardinals learned of the theory and were interested.

Copernicus's challenge was finally published in 1543, as he lay on his deathbed.† Some believe the first published copy was given to him on the

---

* Some seek to expand Copernicus's discovery into something called the "Copernican principle." The Copernican principle claims that Earth is not in a special position, and more generally that there is nothing special about human beings or Earth.

† Copernicus dedicated his masterpiece—*De revolutionibus orbium coelestium (On the Revolutions of the Heavenly Spheres)*—to Pope Paul III. Contrary to popular wisdom, it appears that during his lifetime, Copernicus was never criticized by the Catholic Church or other religious leaders. In the preface, Copernicus argued that mathematics, not physics, should be the basis for understanding and accepting his new theory. Copernicus's theories were criticized by religious leaders after his death. Some say it was a case of "old science" using religion to brush aside "new science." An

day he died. Copernicus shook the ancient paradigm, but it didn't break, at least not completely. More than 140 years passed before the world would generally accept the beauty and simplicity of Copernicus's model. The breakthrough came from a man of perhaps even greater genius.

Isaac Newton

Isaac Newton (1642–1727) is considered by some to have been the greatest scientist to ever live.[3] The book *Men of Mathematics* lists him as one of the three greatest mathematicians of all time.* Newton made major contributions to physics, mathematics, and optics. French mathematician Joseph-Louis Lagrange said Newton was the greatest genius who ever lived, and also "the most fortunate, for we cannot find more than once a system of the world to establish." Newton wrote, "If I have seen further it is by standing on ye sholders of Giants."[4]

Two of those giants were Johannes Kepler (1571–1630) and Galileo Galilei (1564–1642). Kepler described mathematical rules for the orbits of planets. Galileo demonstrated that gravity causes all objects to fall with the same acceleration, regardless of their mass, and with his improved telescope discovered the four major moons of Jupiter.

Newton developed three basic laws of motion.† He added his most astonishing concept—a theory of universal gravitation. Newton proposed that every object is attracted to every other object—instantly and throughout all space—by the force of gravity. He further proposed that the gravitational force between two objects is proportional to the mass of each and inversely proportional to the square of the distance between them. If the distance is doubled, the gravitational force is reduced by a factor of four.

With his breakthrough theory of universal gravitation and his three laws of motion, Newton was able to show mathematically why the

---

interesting book is *The Book Nobody Read—Chasing the Revolutions of Nicholaus Copernicus* by Owen Gingerich (New York: Penguin Books, 2004).

* The other two are Archimedes and Carl Friedrich Gauss. I won this book in a high school math contest, and it made a great impression on me.

† These are: (1) objects in motion remain in motion, unless acted on by a force, and similarly objects at rest remain at rest unless acted on by a force; (2) force is the product of mass times acceleration; and (3) for every action (force) there is an equal and opposite action (force).

planets obeyed Kepler's laws. He also explained the motion of comets, the Moon around Earth, and Earth around the Sun. His theories of gravitation and motion explained why objects remained stationary on Earth as Earth rotated on its axis each day and revolved around the Sun each year.

Newton's principle of universal gravitation and his laws of motion finally tipped the balance in favor of the Copernican model.* The paradigm shifted and Copernicus's model of Earth and other planets revolving around the Sun, all in accordance with Newton's concept of universal gravitation and laws of motion, became the accepted worldview for the scientific community. Today, the Ptolemic model of concentric crystalline spheres around Earth seems ridiculous. But it lasted for over two thousand years, through a large part of recorded human history, and it took hard work by many geniuses to break it. Major scientific paradigms are hard to change. In this case, it took concepts of number— mathematical formulas for motion and gravitation—to change our understanding of the universe.

Newton broke the Aristotelian paradigm a second way. Aristotle thought humanity should be able to figure out the universe through reason alone. Newton showed that experimental testing and observation, and analysis of results, were also needed. After Newton, the modern scientific method gained favor, and the impact of science upon civilization accelerated, slowly at first, but then dramatically.

The basic tenet of the scientific method is simple; if experimental results don't support a theory, the theory must be modified. Yet, even now, experimental facts are typically ignored if they conflict with the existing paradigm of a world created without reason or design.

Newton did agree with Aristotle that the universe was infinite and eternal—that it was unlimited in size and had always existed. Newton also believed in the God of the Bible, an intelligence that created and designed our universe. Newton saw evidence of design everywhere.

---

* The first observational proof of Earth's motion came after Newton's death, with the discovery of a phenomenon called the "aberration of starlight." Because the speed of light is finite, when a body moves transversely to a star, the star's apparent position appears to shift relative to Earth. It is similar to the apparent shifting of the direction of the wind when sailing in opposite directions. Or if you are walking fast in rain that is falling straight down, you may need to tilt the umbrella forward a bit to stay dry.

"This most beautiful system [the universe] could only proceed from the dominion of an intelligent and powerful Being."*

Albert Einstein

A more recent example of a paradigm shift, and one that is central to the ideas of this book, is Albert Einstein's theory of relativity. For over two hundred years, scientists generally thought Isaac Newton had fully explained how the heavens work. Newton's laws of motion and theory of gravity fully explained the orbits of the Moon and the planets, at least within the accuracy of Newton's time. And who could argue with, or doubt, Newton's concepts of absolute space and time? Were they not self-evident? Certainly no established scientist dared challenge the Newtonian paradigm.

But there were some contrary experimental results, some serious "flaws" in the Newtonian worldview. Scientists such as James Clerk Maxwell (1831–1879) and Hendrik Lorentz (1853–1928) developed a theory of electromagnetism that required a fundamentally different view of space and time.

Perhaps the most troubling experiments involved the concept of relative motion. Everyone "knew" that the motion of a body, relative to an observer, depends in part on whether the observer is also moving, and in what direction. Consider a train moving at a constant speed. The rate at which the train approaches an observer standing next to the track is clearly different than the rate at which the train approaches an observer on a different train that is moving. Put another way, if someone is walking toward you, she will reach you faster if you start walking toward her, and she will reach you slower (or not at all) if you start walking away from her. All motion is relative to a fixed framework of absolute space and absolute time, according to the Newtonian view of the world. Any two moving objects were thought to pass through the same immovable grid, so to speak, and any measure of distance and time should be exactly the same according to both.

---

* This quote comes from the "General Scholium," which is an essay Newton appended to his second (1713) edition of the *Philosophiæ Naturalis Principia Mathematica,* more commonly known as the *Principia.*

The problem was light didn't seem to work that way. Scientists imagined that visible light, other radiation, and gravity traveled through "ether," which was thought to be everywhere, to exist throughout space and throughout time. A number of very smart people created experiments to detect the ether. The most famous of these is the Michelson-Morley experiment of 1887. Albert Michelson and Edward Morley figured out a way to detect small variations in the speed of light, using equipment placed on a large block of marble afloat in a vat of liquid mercury. But they couldn't find any. Whichever direction light traveled, whether it was going in the same direction as Earth rotating around the Sun, or against it, the speed of light was exactly the same. There was no evidence of the ether.

From 1790 to 1890, at least four major experiments looked for evidence of the ether. All failed. There was no evidence of any relative motion in the speed of light. All consistently found that the speed of light approaching an observer did not depend in any way on whether the observer was moving. This was contrary to the Newtonian paradigm. No established scientist was able to explain it, perhaps because it was not possible to find a solution within the Newtonian paradigm, and scientists were not willing or not able to think outside the dominant Newtonian paradigm.

The answer came from outside the scientific community of the day, from a Swiss civil servant whose official title was patent clerk third class. One problem the Swiss were trying to solve was how to synchronize clocks in different cities, so that trains could run on time with closer schedules and ships could navigate more efficiently. Various solutions were proposed, and some of their patent applications were reviewed by a young man who had been unable to find an academic position. His name was Albert Einstein (1879–1955).

Einstein's workload was light, at least for him, and he was free to spend a good portion of his days thinking about clocks and the concept of time. He made up a number of thought experiments to help understand the problem. In one of these, he imagined two towers. What does it mean, Einstein asked, to say that they are both struck by lightning at exactly the same instant? He realized that, because the speed of light is finite, whether the lightning strikes were simultaneous, in the eyes of an observer, would depend on where the observer was located. Lightning strikes that were simultaneous to one observer closer to tower 1 would

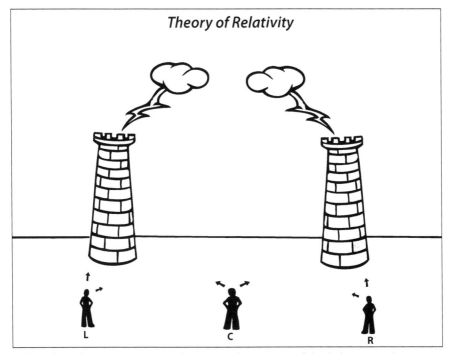

Theory of Relativity

Light takes a longer time to travel a greater distance, so if the lightning strikes appear simultaneous to C, L will observe the tower on the left being struck first, and R will observe the tower on the right being struck first.

not appear simultaneous to another observer in a different location closer to tower 2.

Einstein eventually came to the shocking conclusion that there is no such thing as absolute time. In effect, each observer carries his or her own clock. Each clock is separate, although there are rules for comparing one clock to another, and clocks that are not in relative motion will measure the passage of time in the same way.

With this "freedom," so to speak, Einstein then asked what the rules of our universe would be if the experiments were correct. What if the speed of light really was the same as measured by all observers in different states of motion? This led to his theory of special relativity, which he published in 1905, his "miracle" year. Einstein claimed that absolute space and absolute time are illusions. His theory of special relativity was based on two startling, but conceptually elegant, key assumptions:

1. The speed of light is always the same, whether the observer is at rest or moving in any direction at any speed, and
2. The laws of physics are exactly the same, whether the observer is at rest or moving in any direction at any speed.

In other words, you can't tell the difference—by measuring the speed of light or performing any other experiment—between being stationary or in constant, unaccelerated motion. There is no fixed background "grid" of absolute space and time.

In Einstein's theory of relativity, nothing can move faster than the speed of light. This didn't agree with Newton's theory that the force of gravity acted instantly throughout the universe. Einstein worked hard to develop a theory that included gravity, and in 1915 he succeeded. He created his theory of general relativity by adding a third key assumption:

3. The speed of light, and all other laws of physics, appear the same in a gravitational field as they do in uniform acceleration of the same strength.

In this "general" theory of relativity, gravity is the result of space being curved by matter. As physicist John Archibald Wheeler put it: "Matter tells space how to curve. Space tells matter how to move."

These three key assumptions, or postulates, and the resulting equations, are very elegant in a mathematical sense. It is as if there is a grander scheme at work, hidden symmetries and equivalences between space, time, and the laws of the universe. Einstein used mathematics—in this case powerful and now proven connections between space and time—to change the Newtonian paradigm.

The story of Albert Einstein is a wonderful example of a paradigm shift and of the difficulty in changing how the scientific community views reality. Einstein came from outside the system; he was not a tenured faculty member, and he did not have a personal investment in the existing paradigm. His ideas were not accepted overnight. His theories were not commonly accepted until 1919, after Arthur Eddington was able, during a solar eclipse visible from the island of Príncipe off West Africa, to measure the curvature of light by the Sun. The Sun's gravity "bent" the light from stars behind the Sun, which became visible during the total eclipse, by exactly the amount Einstein predicted. Even with this evidence, some scientists refused to abandon the prior paradigm. As late as 1950, scientists were conducting futile experiments to discover the imagined "ether." Recent experiments have confirmed, to within one part in one hundred million billion (10^{17}), that the speed of light does not change when an observer is in motion.[5]

Tests of general relativity continue. General relativity was able to explain a slight anomaly in the orbit of the planet Mercury, an anomaly that could not be explained by Newton's laws of motion.* An amazing confirmation of general relativity was completed in 2011. Gravity Probe B contained four almost perfectly spherical gyroscopes spinning in liquid nitrogen, in a satellite 400 miles above Earth. The experiment took over fifty years from initial proposal to completion, cost a reported $750 million, and required the development of nineteen new technologies. The gyroscopes confirmed that space is twisted by Earth's rotating gravitational field. Because of the rotation of Earth, the axis of the gyroscopes was altered very slightly over time, by thirty-seven one-thousandths of a second of arc—the equivalent of a human hair ten miles away—each year.†

That kind of experiment may seem far-fetched, but Einstein's theories have practical uses. You probably have a GPS system in your car or phone. That technology uses special and general relativity—it compensates for the relative movement of the satellites and the slowing of time closer to Earth where Earth's gravitation field is stronger.

Albert Einstein created a new paradigm of reality. His concepts of relativity were shocking; they destroyed the Aristotle/Newtonian paradigm of absolute space and time, a paradigm the world had accepted for thousands of years. Despite the obvious elegance and beauty of his ideas, and the repeated experimental confirmation of the fixed speed of light, Einstein's concept of reality took many years to be accepted.

Although he was ultimately hailed as a symbol of scientific genius and a model for the scientific community, and became by far the best-known scientist of his time, Albert Einstein began as an outsider, and that ability

---

* Einstein may not have realized that this anomaly could be explained by general relativity.
† See "52 Years and $750 Million Prove Einstein Was Right," *New York Times,* May 4, 2011.

to think outside of the prevailing scientific paradigm helped establish his greatness. Up to Einstein, scientific thought began with concepts of absolute space and time. No member of the established scientific community was able to break free of that paradigm.

The Bias against Design

Today, vocal parts of the scientific community vigorously reject any notion of design in the universe and any suggestion of a greater reality. Some attack even the most innocent suggestions of design. For examples, see Ben Stein's excellent documentary, *Expelled—No Intelligence Allowed*. I met one of the victims years ago and offered legal help to fight his mistreatment by the Smithsonian Institution.* This strong

---

* Richard Sternberg has two PhD's, one in molecular biology and one in theoretical biology. He was a well-respected researcher for the Smithsonian Institution and the editor of a technical journal called the *Proceedings of the Biological Society of Washington*. He is fair, open-minded, and an accepted scientist with more than thirty peer-reviewed articles in scientific books and publications. He also struck me as one of the nicest, kindest persons you are likely to meet.
Richard Sternberg allowed his journal to publish a scientific article that noted evidence of design in the history of life. He didn't coauthor the article; he simply allowed it to be included in the journal after it had been peer reviewed by three other scientists. The article was by biologist Stephen Meyer and about the sudden appearance of numerous complex forms of life in the so-called Cambrian explosion of around 540 million years ago (we'll come back to this in chapter 12). Meyer's article concludes in part that analysis "suggests purposive or intelligent design as a causally adequate—and perhaps the most causally adequate—explanation for the origin of the complex specified information required to build the Cambrian animals and the novel forms they represent."
All Sternberg did was publish this article in his technical journal after it had been peer reviewed. That simple act provoked a firestorm of indignation, retaliation, and discrimination. He was demoted, given a hostile supervisor, forced to move offices twice and ultimately to use a shared work space, and his research privileges were limited, which is career death to a research scientist. Enormous pressure was placed on him to resign in disgrace. An investigation by congressional staff "uncovered compelling evidence that Dr. Sternberg's civil and constitutional rights were violated by Smithsonian officials." The staff report concluded that "[t]he failure of [Smithsonian officials] to take any action against such discrimination raises serious questions about the Smithsonian's willingness to protect the free speech and civil rights of scientists who may hold dissenting views on topics such as biological evolution."
All of this struck me as terribly wrong. Why should Richard Sternberg not be allowed the freedom to edit his journal and publish articles in his journal as he saw fit? Why should he be attacked by Smithsonian officials and special interest groups merely for allowing a discussion on

bias, this discrimination, this violation of intellectual freedom, left me feeling a little angry, and a little ashamed of the scientific community. What was wrong with discussing the mere possibility of design? Why wouldn't other scientists come to the defense?

Scientific paradigms, accepted ways of understanding the world, are hard to change. Max Planck, a great scientist and a discoverer of quantum physics, offered these words in his autobiography: "A new scientific truth does not triumph by convincing its opponents and making them see the light, but rather because its opponents eventually die, and a new generation grows up that is familiar with it."[6] Some say science progresses one funeral at a time.

You have a choice. You can impartially consider new scientific evidence as it becomes available and revise your opinions as dictated by facts. Or you can do what most people do, just go with the existing paradigm. I'm asking you to have the courage to explore the harder path. I'm asking you to consider the evidence, some of the most sophisticated results of modern science, and make your own decision.

---

the existence of design? I invited him to lunch and was amazed at what a genuinely nice guy he was, and it all struck me as very sad. I was troubled that he seemed all alone, with no defender. I offered my legal services, to represent him against the Smithsonian to preserve his job and his reputation. Richard ultimately decided it was not worth the fight, which probably was the right choice, and he moved on.

CHAPTER 6

The Great Debate

What's new in the debate over the existence of God?

Intelligent design will revolutionize science and our conception of the world.

BILL DEMBSKI

As I BEGAN MY quest, I quickly realized I was following a well-worn path, with an early choice to be made. I could walk the path with eyes of wonder and see evidence of design and purpose everywhere. Or I could walk the path and see the same beauty but sense no purpose. We each choose our own way to view the world, our own paradigm of existence.

What surprised me was not this fundamental choice but the animosity between those who view the world in different ways. In 2005, my path led me back to Boston, and I was not prepared for the intensity of the emotions I encountered.

Fall days in Boston can be magical. The city's colleges and universities pulse with thousands of the world's smartest and most idealistic students, drawn by some of the world's most famous teachers. Cool weather helps focus the mind. You get a sense that any problem susceptible to thought can be solved by the awesome brainpower around you. There is a sense of intelligence.

On November 2, 2005, I went to Boston to hear a debate on "intelligent design." To me, intelligent design is simply the observation that design is a better explanation for the origin of life and the complexity of living systems than natural selection and neo-Darwinian theory. According to Bill Dembski, a leader of the intelligent design revolution and one of the debaters that night, "Simply put, intelligent design is the science that studies signs of intelligence."[1] Intelligent design is based on the latest scientific information about how life really works, including the complexity within each cell in our bodies. Intelligent design uses probability theory and advanced mathematics regarding the complexity of information to conclude it is unlikely, very, very unlikely, that many of the amazing systems found in all life arose solely by natural selection. Chapters 9, 10, and 11 explore these concepts.

Intelligent design has been viciously attacked by the popular press and the National Science Foundation, who claim it is not science. But the real issue is not whether intelligent design meets some contrived definition of "science." (Intelligent design passes Richard Feynman's test of what is science—it is consistent with observation and experiment.) The real issue is whether there is evidence of design in biological systems. Accident or design? That is the question.

This bias against intelligent design was obvious that night. The debate was not about whether intelligent design is or could be a valid observation. The debate was limited to whether a public high school may even tell its students that the concept of intelligent design exists. The title of the debate was simply "Should Public Schools Teach Intelligent Design Along with Evolution?" Can students be told that hundreds of scientists disagree with neo-Darwinian theory?

Even this limited issue was and still is controversial. In the weeks prior to the debate, many of the most widely circulated newspapers and periodicals in the United States, including the *New York Times* and the *Washington Post*, denounced intelligent design. The suggestion that there is a philosophical choice to be made was angrily mocked. The *New Yorker* magazine proclaimed, "Intelligent Design is not science." Now the *New Yorker* is one of my favorite magazines, with often terrific writing (and delightful cartoons), but it was sadly wrong. The *American Heritage Dictionary* defines "science" as "The observation, identification, description, experimental investigation, and theoretical explanation of natural phenomena." Intelligent design is based on experimentation, and on the observation and theoretical explanation of natural phenomena. Regardless of what the media or even the courts say, intelligent design is science. Intelligent design is not Scientism—it is not consistent with a belief system that the natural world can, by itself, explain everything about the natural world, including how the natural world came to exist. Intelligent design is not consistent with the views of the mainstream scientific community and our popular media, who ask us to believe only in "science," as if it were a religion.

All of our scientific progress—every discovery, every invention, every theory—has come from observation, experiment, and reasoning. Observation, experiment, and reasoning are science. Science is not the National Science Foundation, and it is not all the institutions of higher learning. Science simply asks whether a theory is consistent with observation and experiment. Intelligent design passes that test.

As I waited for the debate to begin, I thought of how the debate that night was part of an ancient and fundamental conflict—part of the real "great debate." Is what we see and sense all there is, or is there a greater reality, a greater truth, a greater purpose?

The real great debate—over this great question and ultimate truth—has raged for thousands of years.* The battle is fought on many levels. Its major conflicts span generations, and its skirmishes define our times.

Today the forces fighting the great debate are largely divided into two very different camps. One side believes it all just happened, and there is no greater reality. To them, the natural world is all there is, and there cannot be any greater reality. This is Scientism. Persons who believe in Scientism explain the amazing fine-tuning of the universe as simply our good fortune. They note that if it weren't just right, we wouldn't be here

---

* Here's a quote I love from the conclusion of Immanuel Kant's (1724–1804) book *Critique of Practical Reason,* written in 1788. "Two things fill the mind with ever new and increasing admiration and awe, the more often and steadily reflection is occupied with them: the starry heaven above me and the moral law within me." See http://www.rep.routledge.com/article/ DB047. We will look at the starry heaven in chapters 7 and 12. Because this book focuses on the science of belief, we will not consider the philosophic argument that the existence of a moral law necessitates the existence of God. For that I recommend the classic, C. S. Lewis's *Mere Christianity* (New York: Macmillan, 1943).

to observe it.* Some argue that there must be an infinite number of universes, with different physical laws and constants. We exist in a Goldilocks universe, with everything just right, because we could not exist in a different universe, they say (but see chapter 9, "Problems with the Multiverse").

As for the question of how life began and became so advanced, those who believe in Scientism rally behind the image and theories of Charles Darwin. Life began—they sometimes claim—simply because conditions on Earth billions of years ago were favorable (this is not true, as we will see in chapter 10). Life became so advanced—they argue with great fervor— by the inexorable process of natural selection. The fundamental tenet of their theory of evolution, which is often called "neo-Darwinism" (*neo* is Latin for "new") or "neo-Darwinian theory," is that all of the species that have ever existed, and all of the special abilities and adaptations of those species, arose solely because of accidental mutations and natural selection. The Scientism side totally rejects, sometimes with gentle rebuke and sometimes with mocking derision, any suggestion that a greater being or a greater reality is necessary to explain our universe or anything in it. As Atheist Richard Dawkins put it: "Darwin made it possible to be an intellectually fulfilled atheist."[2]

Of course, the battle lines can be confusing. One can believe in neo-Darwinian theory and still believe in God. But neo-Darwinism is the main weapon that Atheist/Scientism/Naturalism forces use to counter our instinctive awe over the design of living creatures.

On the other side of the great debate is a loose coalition I call "Belief." They believe in something greater. Some call it God. Some do not believe in a God, at least not the God of any traditional religion, but they have a general sense of spirituality and believe in some type of greater reality. On the Belief side, there is room for wonder.

For the last 150 years or so, since Charles Darwin's book *On the Origin of Species* was published in 1859, the Belief side has been losing the great debate in the eyes of many Western institutions and intellectuals. In recent decades, what was a retreat has turned into a rout, at least

---

* This is sometimes called the "anthropic principle." You can read a lot of books and articles that discuss the anthropic principle, but to me they are all pretty useless. You can't prove anything with the anthropic principle.

according to our mass media. Attendance at religious services is no longer expected or commonplace among persons with a college degree; in some circles, it is unusual. Other than at a few religious institutions, it is generally impossible for a person who publicly challenges Darwin's theories to become a tenured professor. The "popular" press, who you might hope if they truly understood the subject would embrace freedom of choice and wonder, has instead viciously mocked proponents of design in the universe. A federal court actually ruled, in *Kitzmiller v. Dover Area School District* in 2005, that it violates the constitution of the United States when a local school board informs high school students that intelligent design is an alternative to Darwin's theory of evolution by natural selection.[3] The judge in the case actually ruled at the request of the American Civil Liberties Union, which somehow totally missed the point in this case that liberty includes the right to hold a dissenting view. As a lawyer, I find it embarrassing that the First Amendment to the Constitution, designed to promote freedom of religion, has been twisted to ban the teaching of legitimate science simply because it points to the existence of God. In effect, the case adopts Scientism as the preferred and protected religion of the United States. It is a compound fracture of the legal system, another illustration of the stranglehold of the neo- Darwinian paradigm.

It's bad in the United States, but it's worse in the United Kingdom, where a new law mandates the teaching of Darwinian evolution as a "comprehensive and coherent scientific theory."[4] To me, that's sad and twisted. Politicians pass a law that purports to tell us what scientific theory we should believe in to explain the wonder of life?

To the popular press and leading universities, the great debate is over. Scientism has vanquished Belief. Anyone or any group who dares suggest otherwise is branded a heretic, and is isolated and condemned. The popular paradigm cannot be questioned.

Scientism is on the rise as we turn away from the wonders of existence. A few centuries ago, we farmed and hunted face-to-face with wonder. We have obscured the heavens with light pollution; we do not realize what we have lost. I once sailed all through a cloudless, moonless night in Canada from Nova Scotia to Prince Edward Island. The Milky Way split the heavens, from horizon to horizon. I steered three stars to the left of the Milky Way.

Powerful institutions—most colleges and universities, newspapers, magazines, and television and movie producers—want you to believe that our universe is meaningless and pointless, a grand system where everything somehow arose by accident and with no purpose or design but somehow, miraculously, gives the appearance of design. Their stature is enhanced by this empty yet popular paradigm. To these powerful institutions, there is no greater reality. Religion is generally tolerated, but only in the sense that people should be free to do whatever makes them feel good and only if there is no pretension that religion or a greater reality have any role in explaining the creation of the universe, life, or human beings.

Yet the great debate is far from over. The battle has entered a new and exciting phase. A counterattack has been launched, a counterattack aimed directly at the foundation of the Scientism fortress. This counterattack is based solely on scientific facts and advanced logic. It does not look to the Bible, the Koran, or any other religious or historical source for justification, yet it profoundly supports the concept of Belief.

By using only science—observation, experimentation, and logic—the counterattack exposes Scientism for what it is, a system of belief. A powerful weapon in the counterattack is the concept of intelligent design. It is so powerful that believers in Scientism continually attempt to misstate it or ignore it.

Intelligent design studies biological systems for signs of intelligence. It does this largely by measuring the probability that biological systems could ever have been formed by accident. When the probability gets very low, intelligent design scientists conclude that the system was designed. Intelligent design stops there; it says nothing about who the designer was or what the designer's motives were. Those great questions are beyond science. Intelligent design claims, "[T]here are natural systems that cannot be adequately explained in terms of undirected natural forces and that in any other circumstance we would attribute to intelligence."[5]

Intelligent design has been viciously attacked, not so much for its claim that design can be detected, and not so much for the mathematical methods it uses, but because it trumps the belief system of those who consider themselves to be our ruling intellectual elite. It trumps Scientism. Intelligent design is widely misrepresented. Let's look at how it works.

The claim that design can be detected is intuitively obvious and used in other areas. Search for Extraterrestrial Intelligence (SETI) scientists have been attempting unsuccessfully for decades to detect design/intelligence in signals coming from outer space. (Intelligent design suggests they are looking in the wrong direction and that the evidence of design is within life.) SETI runs signals from outer space through computers programmed to recognize patterns. In the movie *Contact,* scientists conclude a signal was created by intelligent beings when it identifies in order, through beats and pauses, the prime numbers from 2 to 101.

This type of pattern has what intelligent design scientists call "specified complexity."* It is the signature of intelligence. For "specified complexity" you need three things:

1. A recognizable pattern, or "specified" event. In *Contact*, this is the sequence of prime numbers.
2. The pattern must be "complex," in the sense that it is extremely unlikely to have arisen by chance.
3. There must be no known natural cause.

If a pattern or event satisfies these three conditions, we can reasonably assume we have detected evidence of intelligence, just as in the movie *Contact.* We could be wrong. Theoretically, we might someday discover a natural reason why certain stars give out radio signals that, through beats and pauses, identify in order the prime numbers from 2 to 101. But there is no reason at this time to think that will ever happen. So should SETI scientists ever detect such a sequence of primes, they would conclude it was generated by intelligence.

Archeology and forensics also use our ability to detect specified complexity. It is intuitively obvious. A famous example is finding a watch on

---

* "Specified complexity" has its roots in mainstream science, not in intelligent design. It was first proposed by noted origin-of-life researcher Leslie Orgel. In 1973 Orgel stated: "[L]iving organisms are distinguished by their specified complexity. Crystals are usually taken as the prototypes of simple, well-specified structures, because they consist of a very large number of identical molecules packed together in a uniform way. Lumps of granite or random mixtures of polymers are examples of structures which are complex but not specified. The crystals fail to qualify as living because they lack complexity; the mixtures of polymers fail to qualify because they lack specificity." Leslie E. Orgel, *The Origins of Life: Molecules and Natural Selection* (London: Chapman & Hall, 1973), p. 189.

the side of the road. Would you think the watch was created by random processes or by an intelligent being? Or imagine hiking through a remote wilderness and finding a clearing with your name perfectly spelled out in large rocks. You would know someone had done it, because it is (1) a recognizable pattern, (2) impossibly unlikely to have happened by chance, and (3) not a result of any known natural cause.

Impossibly unlikely events happen all the time. To detect intelligence, you need a recognizable pattern. Intelligent design scientists like Bill Dembski have given this subject a lot of thought and have made it precise.* Here's an example from Dembski's book *The Design Revolution* that I find helpful.

Suppose you flip a coin 1,000 times. Whatever sequence of heads and tails you get, that is an amazingly unlikely pattern. Now suppose someone else flips a coin 1,000 times, but he gets all heads. You'd think, you'd know, it was rigged. Sure, the odds of him getting 1,000 heads in a row are the same, 1 in $2^{1,000}$, as of you getting whatever sequence you flipped, but he got a recognizable pattern.

So how unlikely does a pattern have to be to detect intelligence? Does getting 1,000 heads in a row qualify? Here's where number sharpens the concept of intelligent design. You might think that, even though it's not likely to happen to you, or any of your friends, if everyone on Earth stood around flipping coins, it would likely happen soon. You'd be wrong.

Dembski suggests a lower bound, a "universal probability limit," of 1 in 10^{150}. He gets that by taking the number of protons, neutrons, and electrons in the visible universe (10^{80}), multiplying by the number of seconds since the creation of the universe (about 4 times 10^{17}), and multiplying by 10^{43} units of "Plank time" in each second. A unit of Plank time, 10^{-43} seconds, is theoretically the smallest time measurement that will ever be possible. When you multiply those numbers together, you get about 10^{140}, which Dembski rounds up by more than 1 billion (10^9) to conclude that, in the entire history of the universe, there are far, far less than 10^{150} opportunities for any possible event. Below 1 in 10^{150}, you've used up the "probabilistic resources" of the universe. In other words, if

---

* Technically, specified complexity is an information measure, like the information measure introduced by Claude Shannon in the 1940s.

every proton, neutron, and electron was able to perform an "event" (such as flipping a coin 1,000 times) every 10^{-43} seconds since the beginning of the universe, the odds are less than 1 in a billion that an event with a probability of 1 in 10^{150} will ever occur by chance. Intelligent design claims that if an event with a probability lower than 1 in 10^{150} occurs, and if it has a "specified" pattern, whether in the series of coin tosses or in the usefulness of a particular arrangement of atoms, and if there is no known natural cause, then that event has "specified complexity" and is a signature of intelligence.

Getting 1,000 heads in a row has a probability of 1 in $2^{1,000}$, or about 1 in 10^{301}, far below Dembski's "universal probability limit" of 1 in 10^{150}. So if someone flips a coin and gets a thousand heads in a row, you can conclude it was rigged. Intelligent design uses number in this way to challenge the belief system of Scientism. In later chapters, we will look at biological structures where the odds of formation by chance are far less than Dembski's "universal probability limit."

Some think Bill Dembski's universal probability limit is too low. Some would suggest that below 1 in 10^{50} specified events never happen. But I think Dembski is wise to be conservative and propose his much lower limit. If we want to claim something is unlikely to ever occur by chance in the history of the universe, we need to consider the "probabilistic resources" of the entire universe.

Probability and Religion

Probability arguments may seem a long way from religion and from the great debate over the existence of God. To me, they are compelling. Here's a true story of a low-probability event.

The year is 701 BCE. At this moment in history, the Jewish nation is about to be annihilated. The Jewish king Hezekiah is trapped in Jerusalem like a bird in a cage. His city is surrounded by the Assyrian army. It isn't a fair fight; the Assyrian army is unbeatable. On this military campaign alone, they have captured twenty-three walled cities. Jerusalem is next. The empires of Egypt and Assyria are at war. Jerusalem is located between the two empires in a dangerous position and vulnerable to the armies of both. King Hezekiah has sided with the Egyptians in this struggle, and in 701 BCE that appears to be a huge mistake.

Twenty years earlier, another Assyrian army came within miles of Jerusalem. They laid siege to the city of Samaria. If you've never heard of Samaria, it may be because the Assyrians wiped it off the map. Samaria was much larger than Jerusalem, and only about 35 miles away.[6] Samaria was the capital of the northern tribes of the Jewish state. It held out against the Assyrian siege for almost three years.

The lost tribes of Israel didn't take a wrong turn in the desert. They were destroyed by the Assyrians and their king, Sargon II. Some individuals may have survived and intermarried. But as a culture, as a social structure, they were destroyed. The Assyrians brought in new populations to settle the area. The new residents of Samaria were outcasts in Jewish society. Their outcast status is part of the New Testament parable of the Good Samaritan.

During the prior Assyrian invasion, King Hezekiah paid a large tribute, and the Assyrians spared Jerusalem. But in 701 BCE they are back, and they want the city. The Bible records the taunts of the Assyrian general: "On what are you basing your confidence, that you remain in Jerusalem under siege? . . . Do you not know what I and my predecessors have done to all the peoples of the other lands? Were the gods of those nations ever able to deliver their land from my hand? Who of all the gods of these nations that my predecessors destroyed has been able to save his people from me?"[7]

Then something happened. Some think it was plague. Centuries later, Roman historian Herodotus wrote that the Assyrian army was "overrun with rodents." The Bible says God sent an angel who killed the Assyrian army. A more recent theory is a nearby Egyptian army rescued Jerusalem, although that possibility is not supported by the meager historical record. All we know for sure is that the siege was lifted and the Jewish state survived.

Suppose Herodotus got it right, and plague struck the Assyrian army. That would still be an amazingly fortuitous event. Could it have been the hand of God? Perhaps. Does this amazing event, which changed the history of the world, prove there is a God? Of course not. As a matter of logic, it does not prove that God exists. It could have happened by chance. But how fortuitous that plague should break out at that critical moment in history.

As a matter of logic, no event or fact can prove the existence of God, as long as there is any other possible explanation. Similarly, no event or fact can prove God does not exist. But this story of the Assyrian siege of Jerusalem leads to interesting questions. What if it could be proven that our universe is fine-tuned for life? What if it could be proven, almost to a mathematical certainty, that even the simplest form of life could not have arisen by chance? What would the odds have to be for you to believe in God?

The next chapters describe new and amazing scientific evidence of wonder, what I call the science of belief. From cosmology to DNA to quantum physics, it's all there. Let's begin with a great question: Why does anything exist?

PART 2

The Science of Belief

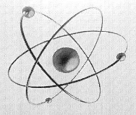

CHAPTER 7

Creation

Why does anything exist?

In the beginning, God created the heavens and the Earth.

GENESIS 1:1

It all started with a Big Bang.

Introduction to *The Big Bang Theory,*
a popular television show on CBS

MODERN SCIENCE SUPPORTS the existence of God in seven ways. Science indicates our universe was created; it has not always existed. To me, evidence of the creation of the universe is the first wonder of modern science, the first of seven in our count to God.

Arno Penzias and Robert Wilson had a problem. They were trying to measure radiation from our galaxy, the Milky Way. The two physicists, working for Bell Laboratories in New Jersey in 1965, had obtained an extremely sensitive microwave antenna. The antenna was twenty feet long and shaped like a large horn of plenty. It was cooled by liquid helium to just 2.7 degrees Celsius above absolute zero, so that it could detect extremely low-temperature radiation.

Penzias and Wilson wanted to measure galactic radiation with a wavelength of about twenty-one centimeters. To make sure the antenna was working properly, they first measured radiation at a smaller wavelength of seven centimeters. They expected to find no radiation at that wavelength. But there was radiation coming from somewhere. For several months they tried to eliminate the "noise." They chased away pigeons and cleaned droppings out of the antenna, which helped a little. But "noise" remained. In addition, and this was particularly strange, the noise—this unknown radiation—was coming from everywhere. It was not affected by Earth's atmosphere. It was not coming from any source on Earth, and it was not coming from the Milky Way or any other particular stellar object or area. It did not vary as Earth rotated or as the seasons passed.

There was no known scientific explanation for radiation at that frequency. Even more puzzling, in the history of science up to then, no one had ever observed radiation coming from all directions. Radiation always had a source, a point or region of emission. Sometimes radiation was observed to reflect off an object or a surface, like light from a mirror, but it had never been found to come equally from all directions. There was no known scientific explanation for radiation that could come from everywhere at once. What could it be?

Penzias eventually posed that question to Robert Dicke of Princeton.* Just the year before, Dicke predicted there should be radiation left over from the birth of the universe, and he proposed an experiment to detect it.† Dicke was sitting in his office with colleagues when Penzias called. It is reported that, when the call was over, Dicke said: "Well boys, we've been scooped."[1]

Ultimately, all agreed this unwanted, irritating "noise" was the greatest prize of all. It was a signal, a relic, from the creation of the universe. It is now called "cosmic microwave background radiation." The photons that make up this radiation are called "relic photons." These photons

---

* Penzias first asked Bernard Burke of MIT, whom he met on an airplane a few months earlier. Burke suggested he call Bob Dicke at Princeton. Just a week before, a colleague of Burke's attended a lecture by Jim Peebles, a scientist at Princeton whose colleague and mentor was Robert Dicke. Dicke had designed (while working earlier at MIT) the microwave antenna Penzias and Wilson were using—called a "Dicke radiometer". The lecture was about a possible experiment to detect radiation from the birth of the universe.

† Actually, this was first predicted by George Gamow in the 1940s. In 1948 Gamow also suggested correctly how hydrogen and helium were created at the beginning of the universe. (Gamow was a bit of an odd duck; he did a lot of his work in the bathtub.) Dicke apparently was not aware of Gamow's earlier prediction.

witnessed the birth of the universe. They have been traveling to us for 14 billion years.*

Penzias and Wilson accidentally discovered the scientific equivalent of the Holy Grail—verifiable proof that our universe began in a single moment. They were awarded the Nobel Prize in Physics in 1978. It may be the most amazing accidental scientific discovery of all time. There is other evidence of creation, some of which is described briefly later in this chapter, and today it is uniformly accepted that our universe had a beginning. But with the accidental discovery of relic photons by Penzias and Wilson, a profound scientific and philosophical debate that had raged for thousands of years began to come to an end.†

Aristotle (384 BCE to 322 BCE) believed the universe was infinite and eternal—limitless in size and had always existed. His paradigm continued for well over two thousand years. It continued through Galileo and Copernicus, through Newton and many other great scientists. These later geniuses altered the paradigm that Earth is the center of the universe but not the paradigm that the universe is infinite and has always existed.

For centuries, the strongest dissent came from the followers of Abraham—Jews, Christians, and Muslims. The Hebrew Bible begins: "In the beginning God created the heavens and the Earth." To followers of Abraham, the universe was created.

Albert Einstein originally believed in an eternal universe. After 1915, when Einstein proposed his general theory of relativity, other scientists pointed out that his equations suggested a universe in motion, either

---

* In the very beginning, the universe was too hot for light to travel freely. Relic photons were all emitted about 380,000 years after creation, when the universe cooled below about 3,000 degrees Kelvin, or about 5,000 degrees Fahrenheit.

† Before leaving Penzias and Wilson, we might applaud the contributions of their employer: Bell Laboratories, the research arm of what was then the phone company—AT&T—to pure science and our modern world. Penzias and Wilson worked at the main location of Bell Labs in Murray Hill, New Jersey, about twenty miles west of Wall Street in Manhattan. Thirteen scientists have shared seven Nobel Prizes for their work at Bell Labs. Bell Lab scientists invented the transistor, which is the key active component in most modern electronics, including computers. Bell Lab scientists invented the laser, radio astronomy, and the UNIX computer operating system, and, in 1926 they created the first motion picture with synchronized sound. Not bad for a private company. So every time you go to the movies, or use a computer or a cell phone, you can thank Bell Labs (or when you have computer problems or get an unwanted call, you can perhaps blame them).

expanding or collapsing. Einstein did not want to accept that. To accommodate the prevailing scientific view at that time that the universe was eternal, he fudged his equations. He added a "cosmological constant" to keep the universe in balance. Einstein later called this cosmological constant the biggest mistake of his life.*

In 1914, the American astronomer Vesto Slipher discovered that almost all of the fuzzy objects in the sky were moving away from us. Even though he didn't know what the fuzzy objects were, he was able to measure their movement by studying the light they gave off. The light was notably weaker at certain frequencies corresponding to known atomic "absorption lines." For example, the simplest element, hydrogen, with just one electron, has an absorption line at a wavelength of 656 nanometers (a nanometer is a billionth of a meter), called the H-alpha line. In the light from stars, these photons are typically absorbed when a hydrogen electron changes energy levels.

For almost all of these fuzzy objects, the frequency of the H-alpha and other absorption lines was shifted slightly. The absorption lines appeared at a lower frequency (which means a longer wavelength) than what would be measured in a laboratory on Earth. For each fuzzy object, the absorption lines were shifted—meaning the light waves were stretched—by similar proportions.

To a scientist, the stretching of wavelengths has a clear meaning. It means that the fuzzy objects—whatever they are—are moving away from us fast. This is the "Doppler effect." When two objects are moving apart, wavelengths are stretched and the frequency of the waves is reduced. It also applies to sound waves. We notice a change in pitch when a siren or a train moves past us. When a siren is coming toward us, we hear a higher pitch than when it is moving away. Movement of the source toward us

---

* In 1998, two teams of scientists independently came to the same shocking conclusion: the expansion of the universe is accelerating. This discovery has resurrected the cosmological constant, but in an opposite way from that first conceived by Einstein. Rather than keeping the universe static, the cosmological constant now represents an unknown force splitting the universe apart. This unknown force is now called "dark energy." Dark energy is believed to make up about 73 percent of the total amount of mass/energy combined in the universe. For this calculation, mass is converted to energy (or vice versa) using Einstein's famous equation $E = mc^2$.

compresses the sound waves and increases their frequency or pitch, and movement away stretches out the sound waves and reduces their frequency or pitch.

Light travels incredibly fast, over 186,000 miles a second. These fuzzy objects had to be moving away at a very high speed for the effect to be measured. But what were they?

On April 26, 1920, in the Baird Auditorium of the Smithsonian Museum of Natural History in Washington, DC, two scientists debated the nature of these fuzzy objects. Their names were Harlow Shapley and Heber Curtis, and their debate that day is still often referred to as astronomy's "great debate." Shapley argued that our Milky Way Galaxy was the entire universe and that the fuzzy objects were part of the Milky Way. He pointed out that a nova in the fuzzy object we now call the Andromeda Galaxy had for a while been brighter than the entire rest of that object combined, and argued that a single star could not appear that bright if the fuzzy object was a far-off separate galaxy. (We now know that this nova was a supernova, which amazingly can outshine for a while an entire galaxy of hundreds of billions of stars.) Curtis discussed the redshifts in the absorption lines of the fuzzy objects and argued that they could not be moving away from us so fast and still be in our galaxy. Both made numerous other points and arguments, submitted opposing papers before the debate, and then wrote further rebuttal papers the year after the debate. As we will see, Curtis was ultimately proved correct on the key point over the "scale of the universe," that the Milky Way Galaxy was not the entire universe, but Shapley was ultimately proved correct on a second important point in the debate, the Sun's position within the Milky Way. Shapley was correct that our Sun is far from the center, whereas Curtis placed the Sun near the center of a smaller Milky Way.

Slipher used a telescope with a 24-inch mirror when he made his redshift discovery in 1914. In the 1920s, using a telescope with a 100-inch mirror, which was about sixteen times more powerful,* Edwin Hubble realized those fuzzy objects were not just gas clouds. He discovered that they contained billions of stars—they were entire galaxies like our own Milky Way. It was curious that almost all of these massive collections of

---

* Roughly speaking, the ability of a telescope to detect faint objects depends on how much light it collects. A 100-inch telescope collects over sixteen times the light of a 24-inch telescope (the surface area of the mirror is over sixteen times larger), and thus in a general sense is over sixteen times more powerful.

billions of stars (our Milky Way Galaxy is believed to have 200 to 400 billion stars) were moving away from us, receding, at very high speeds—typically hundreds of miles per second. What could cause that?

One of the great challenges facing astronomers over the past one hundred years has been the calculation of distances in space. How far away are the stars and these other galaxies, these other island universes? For nearby stars, as Earth revolves around the Sun, a "wobble" can be detected and measured as the nearby star appears to shift in position compared to more distant stars. This is a cute trick, but it obviously does not work for distant objects, and it certainly does not hold for far-off galaxies.

A variety of techniques were invented and applied. A key breakthrough was the discovery that for certain variable stars—stars that varied in brightness in a fixed way, like clockwork—one could determine their intrinsic brightness. Their intrinsic brightness depended on the period of the oscillation of their light. So when one of these unusual stars was found, one of these "standard candles," astronomers could measure how bright it appeared on Earth and estimate its distance. These unusual pulsating stars are called "Cepheid variables"; they have five to twenty times the mass of our Sun and are up to thirty thousand times as bright.* Over seven hundred Cepheid variables have been identified in our Milky Way, and they have been detected in galaxies up to 100 million light-years away.

Hubble and others found some of these variable stars, these "standard candles," in what they now knew were far-off separate galaxies. They began the difficult process of estimating distance. By 1929, Hubble had estimated the distance to twenty-four nearby galaxies. He created a graph with distance from Earth on the horizontal axis and speed away from Earth (as measured by the change in frequency, or redshift, of the light) on the vertical axis. For each of these twenty-four galaxies, he marked their distance/speed as a point on the graph. The result is the famous graph on the next page.

Hubble discovered that the distance away from us was generally proportional to the redshift, or the speed away from us. In other words, galaxies twice as far away as others are moving away from us about twice

---

* Hubble discovered Cepheid variable stars in the Andromeda Galaxy (M31) and calculated the distance to that galaxy as greater than the size of the Milky Way estimated by Shapley.

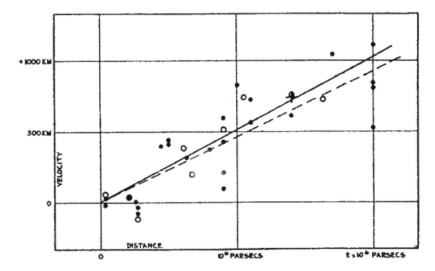

Hubble's chart of galactic redshifts and distances. Reproduced courtesy of NASA.

as fast, galaxies four times further away are moving away about four times as fast, and so on. This astonishing relationship between distance and receding velocity is generally true throughout the entire universe.* It is now called Hubble's law.†

---

* A notable exception is our "close" neighbor the Andromeda Galaxy, which is another island universe similar in size and structure to our Milky Way Galaxy and the most distant object visible to the unaided eye, with an estimated 1 trillion stars. The Andromeda Galaxy is our neighbor, and it is "only" about 2.5 million light-years away, or about 15 million trillion miles away. Because Andromeda and the Milky Way are relatively close, gravity has overcome the general expansion of the universe. The two galaxies are heading toward each other at around 70 miles per second, or roughly 2 billion miles a year. In about 4 billion years, these massive galaxies will "collide" (their relative motion toward each other will speed up as the galaxies get closer). However, because the individual stars are so far apart, it is unlikely that even a single star will literally collide with another.
† I find it interesting, almost humorous, that Hubble was significantly off in his calculations of distances. The slope of Hubble's graph is now called the Hubble constant, and it is generally stated in kilometers per second per megaparsec. A megaparsec is about 3.26 million light-years, and a light-year is about 6 trillion miles (6,000,000,000,000 miles) or 10 trillion kilometers. Hubble estimated this constant, the slope of his graph, to be about 500 kilometers per second per megaparsec. With this original value, a galaxy 10 megaparsecs away—some 32.6 million light-years—would be receding at 5,000 kilometers per second. Measurements in 2010, using the telescope in space named after him, the Hubble Space Telescope, and taking into account the

Hubble's law has religious implications. If you imagine the universe billions of years earlier—you play the movie of time "backward" so that all the galaxies are coming together—it suggests the universe began in a single moment of creation. Hubble's law suggests that, if you reverse time, all of the galaxies might collide at about the same time in the distant past. Because galaxies twice as far away as others would then be heading back twice as fast (if we imagine time going backward), they could all collide at or about the same time and at or about the same place in the distant past. Hubble's law suggests that the universe had a beginning, that the universe was created, as claimed by the faiths of Abraham. The original estimate of when creation took place was between 10 and 20 billion years ago. In Hubble's law, science took an important step toward supporting a religious belief and rejecting prior beliefs in an eternal universe. Einstein visited Hubble after his discovery and said: "I now see the necessity for a beginning."

Hubble's discovery sparked disbelief, sarcasm, and controversy. One famous and then Atheist astronomer, Fred Hoyle, mockingly referred to the implied concept of creation as the "big bang" theory. The name stuck. Despite Hubble's evidence, a competing theory, called the "steady state" theory, was widely accepted. As its name suggests, according to the steady state theory, the universe remains relatively constant. To explain away Hubble's findings, it was claimed that, as the galaxies flew apart, new matter was created to maintain the same overall density of matter in large regions of space. Keep in mind that there was absolutely no scientific evidence that such new matter had ever been created. There was no scientific theory that allowed the creation of matter from nothing—in fact, a basic law of science, the first law of thermodynamics, is that the total amount of matter and energy in the universe remains constant. The steady state theory was bad science. Its sole purpose was to avoid the obvious philosophical and religious implications of the big bang. Its advocates started with a firm conviction that the universe was eternal,

---

bending of light by gravity as predicted by Einstein, found the Hubble constant to be 70.6 plus or minus 3.1 kilometers per second per megaparsec. With this latest value, a galaxy 10 megaparsecs away is receding at about 706 kilometers per second, instead of 5,000 kilometers per hour as predicted by Hubble. Hubble got it wrong by a factor of seven, and he still got his name on the space telescope.

and they desperately tried to invent an antidote to Hubble's law, even at the cost of violating basic principles of science.

The 1965 discovery by Penzias and Wilson of background radiation from the big bang—relic photons emitted 14 billion years ago—killed the steady state theory. Today, at this dawn of the third millennium, scientific evidence of creation is overwhelming. The spectrum of the radiation fits perfectly with the expected "blackbody" radiation from such an event. According to one scientist, it "tells us that the universe was once so dense that it was a single, continuous body in thermal equilibrium, one that could be characterized by a single temperature."[2]

At first, the radiation was thought to be too uniform, and an absolutely uniform background radiation could not explain how matter had condensed into galaxies. However, 1992 measurements revealed very slight variations, on the order of 1 part in 100,000, that closely match scientific models of the formation of galaxies and galaxy clusters. Nuclear physicists have shown that the amount of certain elements, such as helium and deuterium (hydrogen with an extra neutron), closely match their calculations of what happened during the creation of the universe. Even the hottest stars do not "create" hydrogen. Stars like our Sun "burn" by converting hydrogen into helium. In stellar explosions—such as supernovae—much greater temperatures combine hydrogen and helium to forge oxygen, nitrogen, carbon, and other elements necessary for life. The measured amounts of different elements in the universe closely match calculations by nuclear physicists of how our universe was created and has evolved.

Some scientists now believe they know how the big bang unfolded going back to a very small fraction of a second—less than a trillionth of a trillionth of a second—after the big bang. This seems fantastic, and it's hard to see how they can be confident of the physics at such almost unimaginably high pressures and temperatures that perhaps will never be recreated, but it demonstrates enormous confidence in the big bang theory. Scientists now believe the background radiation discovered by Penzias and Wilson was released 380,000 years after the big bang, when temperatures had cooled enough to permit radiation to travel freely. These relic photons were produced when the expanding universe cooled below about 3,000 degrees Kelvin, or about 5,000 degrees Fahrenheit. They have been weakened—redshifted—about a thousand times as the universe has expanded that much in the 14 billion years they have been traveling.

There is one other piece of strong scientific evidence that our universe was created. Thus far, we have Hubble's law—the universe is expanding and more distant galaxies are moving away more rapidly. We have the discovery of actual relic photons from the big bang. The spectrum of that radiation, and very slight variations in the intensity of that radiation, closely match scientific models of an initial explosion and the later condensation of matter into stars and galaxies. Other scientific predictions of a created universe closely match the observed relative amounts of hydrogen and deuterium, elements that could not have been created later inside stars.

The last piece of scientific evidence is almost too simple. You can do the experiment yourself; you can actually detect strong evidence that our universe is not infinite and eternal. Go outside at night and look at the sky. Is it dark, or is it as bright as the Sun?

Why is the night sky dark? It's an ancient and profound question. A dark night sky is not consistent with an infinite and eternal universe. As Edgar Allen Poe stated:

> Were the succession of stars endless, then the background of the sky would present us a uniform luminosity, like that displayed by the Galaxy—since there could be absolutely no point, in all that background, at which would not exist a star.

In other words, if the universe were infinite and eternal, every line of sight would end on the surface of a star. It would be like standing in the middle of a very large canopied forest—in all directions one would see the trunk of a tree and nothing else. You can show this mathematically by thinking of Earth as surrounded by an endless succession of cosmic "shells" containing stars and galaxies. If the universe were infinite and eternal, and assuming that stars and galaxies continue throughout such a universe,* then each "shell" would contribute roughly the same amount

---

* Modern cosmology has generally assumed that on very large scales, say, a billion light–years or more, stars and galaxies are distributed fairly uniformly throughout the universe. It's not clear that this is true. Some scientists recently claimed to have discovered a collection of quasars 4 billion light-years long, which would be by far the largest-known structure in the universe. See Jacob Aron, "Largest Structure Challenges Einstein's Smooth Cosmos," *New Scientist,* January 11, 2013, http://www.newscientist.com/article/dn23074-largest-structure-challenges-einsteins-smooth-cosmos.html. However, even if there are large structures, this obviously doesn't

of light. As you consider more and more of these endless "shells," the amount of light reaching Earth grows until light is coming from every point in the sky.

The riddle of the night sky has been called "Olbers's paradox." Heinrich Wilhelm Olbers was an amateur German astronomer who described this apparent contradiction in 1823, but he was not the first to do so. The contradiction between the dark night sky and the concept of an infinite and eternal universe was noted by Johann Kepler in 1610. Of course, Olbers's paradox is not really a paradox. It is scientific confirmation—a combination of logic and observation—that our universe was created, as claimed by the faiths of Abraham.

The big bang created the entire universe in one magnificent event. It created space and time, and all matter and energy. It did not occur in a specific part of the universe; rather, it occurred simultaneously everywhere in the universe. The universe began as a point. What we see from Earth— galaxies moving away from us in accordance with Hubble's law, so that the farthest galaxies are moving away the fastest—is happening all over the universe. Space itself is expanding.

One way to get a sense of how our three-dimensional space can be expanding everywhere in the universe is to go down one dimension and consider a two-dimensional surface. A two-dimensional surface is like a piece of paper; you can move in two perpendicular directions. (Our three-dimensional space has three perpendicular dimensions.) Now imagine this two-dimensional surface curved into itself, like the surface of a balloon, and imagine the galaxies as dots on the surface of that balloon. As the balloon expands, all the dots move apart from each other, and the dots furthest away move away the most. If a two-dimensional being on the surface of the balloon is not aware of the curvature of the surface of the balloon—his "third" dimension—then he observes his own version of Hubble's law as the balloon expands: the dots/galaxies twice as distant

---

change the mathematical proof described above. The only assumption one would need to make about an infinite and eternal universe is that there is some minimum distribution of stars/galaxies through the universe—in other words, that such a universe does not become empty as you go further out.

are receding twice as fast. Our three-dimensional space is expanding in the same sort of way, moving apart everywhere.

Today, the big bang theory is the accepted model for the creation of the universe. Scientists have concluded that the big bang occurred almost 14 billion years ago. Prior to that date, our entire universe did not exist. Religious belief in a universe that was created, rather than a universe without a beginning, has become accepted scientific theory. Time and space, and all of the matter and energy that ever has existed and ever will exist, were created in a single instant from absolute nothingness, an event that cannot and never will be explained by the natural laws of what now exists. If that is not scientific evidence of wonder, then I do not know what is.

It took a few minutes to cool down so atoms could form, it took another few hundred thousand years to cool down more so light could travel, and it took almost another 14 billion years to get to the universe we see today. But it was all created at the very beginning. It all started with a big bang.

The First Cause

What caused the big bang? What caused our universe to exist? In the sixth century CE, a group of Arab intellectuals argued that the universe had to have been created and that a created universe implied the existence of God.

Their argument is now known as the Kalam cosmological argument. The name "Kalam" comes from a tradition of Arab intellectuals. The argument goes like this:

1. Anything that begins to exist has a cause.
2. The universe began to exist.
3. Therefore, the universe has a cause.

Part 1 above is philosophic. We will discuss it below, but it is hard to deny. Part 2 above is the big bang, now the accepted scientific theory of creation. Part 3 follows from parts 1 and 2.

You either admit something caused the universe to exist or you have to believe things can begin to exist with absolutely no cause. Keep in mind science has never found anything that begins to exist that does not have a cause. Things do not just pop into existence, with no cause and no explanation. This has been true for all of human history. When you lock the door to your house, you do not expect to come back and find strange objects inside—and if you ever did encounter such an object, you would quickly look for a cause. We may not always know the cause of an event, but a fundamental premise of science is that all events have causes, even if you can't see or directly detect them.

Assuming then that the universe had a cause, we are forced to accept that some being or some thing outside of our universe caused it to come into being. Something that does not exist in our reality created our reality. This is a fundamental tenet of the faiths of Abraham. It is also, in my view, an inescapable logical and scientific conclusion. Abraham's belief that something outside of our universe (our reality) caused our universe (reality) to come into being now follows from a simple combination of logic and scientific theory. We have not established that there is a god, and we certainly have not proven that the God of the Bible exists, but we have established that there is something outside of our universe, and that unknown something caused our universe to exist. That is an important start.

Those who are somewhat familiar with quantum physics may object at this point. One feature of quantum physics is that field fluctuations can come into existence and then disappear. These are called "virtual particles," and they are believed to exist, although generally for very short periods, such as a thousandth of a billionth of a billionth of a second.* Perhaps, one might argue, quantum fluctuations caused our universe to come into being.

The clear flaw in this reasoning is that virtual particles do not come from nothing. Virtual particles come from a quantum field that permeates all of space and time—our entire universe. The quantum field is a high-energy field. Put another way, before you can have a virtual particle, you first have to create space and time, and the quantum field, as was done in the big bang. Prior to the big bang, space and time did not exist, and there was no quantum field. So quantum effects in our universe could not have caused the big bang, because there was no time or space, or quantum field, before the big bang.

---

* Virtual particles are believed to exist for about 10^{-21} seconds. This is a very small period of time. During this period, light crosses less than one hundredth of the radius of a hydrogen atom.

To me, it's sort of like the joke where scientists are trying to convince God they can do anything God can do. "Like what?" asks God. "Like creating human beings," say the scientists. "Show me," says God. The scientists say, "Well, we start with some dust and then . . ." God interrupts, "Wait a second. Get your own dust."[3] So if anyone wants to claim our universe popped out of a quantum field, they need to explain what caused the quantum field to exist.

One of the basic tenets of Scientism is that nothing can happen outside of the laws of physics—that miracles cannot happen. The big bang is a clear contradiction.

And so we come to the last stand of diehard Atheists, the Anti-Faith side of the great debate. By now, you should not be surprised to learn that this Atheist position is also today's accepted scientific paradigm. It is believed that there is some sort of universe-creating mechanism that somehow spews out an infinite number of universes with different physical constants. This infinite collection of universes is often called the "multiverse." The existence of an infinite number of universes— of a multiverse—is necessary under this Anti-Faith approach to avoid two serious problems. First, if there were only a finite number of other universes, then there must have been a first universe (or set of first universes) that came into being. What then was the first cause of the first universe (or set of first universes)? Only by believing in an infinite number of other universes, universes without number and without a beginning, can the Anti-Faith believer avoid the profound theological implications of a first cause.

The second reason why the Anti-Faith side believes in an infinite number of universes is to avoid the amazing coincidences discussed in the next chapter. It is scientific fact that our universe is made exactly right, fine-tuned so to speak, for life. We will see how the laws of nature appear to have been arranged, and dozens of factors appear to have been exactly set, so that life may exist. With different laws, and even very slightly different "settings" of those laws, life could not possibly exist. One has to believe in an infinite number of universes, with every possible sort of different natural laws and constants, to explain away the apparent perfect design of our universe. One also has to believe that there is some unknown principle or area of physics that could generate universes with different laws and constants.

72

So we basically come down to two possible explanations for the big bang. The first explanation, proposed by the faiths of Abraham,* is that something outside of our reality created our universe. The second explanation is that there is an unknown branch of physics, of which—if one is intellectually honest—modern science does not have a clue, which somehow creates an infinite number of universes. You can choose to believe either one. Both require you to believe that something outside of our universe created our universe. In other words, both require you to believe in a greater reality. Scientific evidence of the big bang—the creation of space, time, matter, and energy from a single point—forces you to confront the question of first cause.

The creation of the universe does not get one automatically to a belief in God, but it does get you close. The big bang has turned the tables on those who doubt the existence of God. Originally, when the prevailing paradigm was that the universe was infinite and eternal, followers of the faiths of Abraham had an uphill argument—they had to argue that the universe was created and that the first cause must have been God. Now, the tables are turned. Religious belief in creation now fits squarely within the evidence of modern science and prevailing scientific theory. Creation implies a first cause; something outside of our universe that caused our universe to exist.

Chapter 9 is titled "Problems with the Multiverse" and describes why, to me at least, it is easier to believe in God than to believe in the multiverse. For now, let's just note that, to explain away the scientific fact of creation and the amazing coincidences described in the next chapter, Anti-Faith adherents have to believe in three things:

1. There is some unknown branch of physics capable of producing universes with laws and constants of physics.
2. This unknown branch of physics has actually produced an infinite number of universes, without beginning and without end.
3. Somehow, all of this just exists for no reason and without cause.

These three beliefs, which are not supported by scientific evidence, underlie the concept of a multiverse. To many who view existence as

---

* Some scholars believe this concept traces back to at least the Babylonian captivity, when parts of the book of Genesis may have been written.

without meaning, these are reasonable beliefs, and the alternate concept of a creator God is literally unthinkable. To many who embrace the Abrahamic faiths, the concept of a creator God is comfortable and natural, and these three beliefs are unnatural, perhaps comical. The paradigm clash is huge, with a vast chasm between opposing sides of the great debate.

Of course, a person can believe in the multiverse and still believe in the God of the Bible.* God could have created more than one universe. But I think most people who think the multiverse really exists are Atheists.

I do not fault those who choose to believe in the multiverse. I simply assert that belief in a creator God is at least equally reasonable. To me, it is now the Anti-Faith adherents who have the weak position, who argue against the evidence of modern science.

And this is only the beginning. If the only scientific evidence for God were creation, then God would be viewed by some as only a name for the mystery of the big bang. As we will see, the hypothesis of God, the theory of a creator and designer, has six other scientific pillars of support. We have only begun to count to God.

---

* This has been argued by philosopher John Leslie and cosmologist Don Page, for example.

CHAPTER 8

Fine-Tuning

How can it all be so perfect?

A common sense interpretation of the facts suggests that a superintellect has monkeyed with physics, as well as with chemistry and biology, and that there are no blind forces worth speaking about in nature. The numbers one calculates from the facts seem to me so overwhelming as to put this conclusion almost beyond question.

FRED HOYLE

For this is what the Lord says—He who created the heavens, He is God; He who fashioned and made the Earth, He founded it; He did not create it to be empty; but formed it to be inhabited.

ISAIAH 45:18

IT IS NOW accepted, I think by all scientists, that if many features of our universe were just slightly different, life could not exist, and that the universe at least gives the appearance of having been designed with almost unimaginable precision. To me, the discoveries of the fine-tuning of the universe are a wonder, the second of seven in our count to God.

If you've ever taken a course in physics, you know it is heavy in math. There are many equations, and some of them have fixed numbers—constants—put in to make those equations work. These constants of physics have been measured by experiments. One important constant is the gravitational constant, which appears in Newton's law of universal gravitation and in Einstein's theory of general relativity. The gravitational constant measures the strength of the gravitational force.

Gravity is much, much weaker than the other three fundamental forces in the universe,* but it does have one very important thing going for it. It always attracts. And given an object big enough, such as Earth or the Sun or even the Moon, gravity easily dominates. On cosmic scales, gravity is critical. Gravity causes galaxies, stars, and planets to form.

Is gravity strong enough to end the current expansion of the universe and cause it to collapse? Or is the expansion of the universe overcoming and weakening the universe's gravitational attraction? The point at which the two factors are in balance is called the "critical" density of the universe. The ratio of the universe's actual density to the critical density is generally called "omega"—Ω—the last letter of the Greek alphabet.

A current calculation puts Ω within 4 percent of one.[1] Gravity and the expansion of the universe are roughly in balance. The universe appears "flat," not "open" in the sense of runaway expansion or "closed" in the sense of ultimate collapse. It seems odd that this "flatness" would exist, but the really amazing part is that conditions had to be set exactly right in the big bang for this to happen. If the critical density were very slightly less than one at the big bang, then Ω would be small today because of runaway expansion. If the critical density were very slightly more than one at the big bang, then Ω would be large today because of gravitational collapse. It's like shooting a rocket from Earth so precisely that it hangs in Earth's gravitational field for billions of years, on the edge between falling back to Earth and escaping to space.

In 1982 Paul Davies wrote *The Accidental Universe.*[2] A key theme was how sensitive the universe is to certain key factors. Davies wrote, "[I]t is

---

* There are four basic forces in the universe. Two of these—the "strong" nuclear force and the "weak" nuclear force—are very short range; they operate in the nuclei of atoms. Electricity, or electromagnetism, is a third force. The electrical force basically drives all chemical and biological actions and reactions. We walk, we talk, we think, using the electrical force. The fourth force is gravity. Gravity is much weaker than the electrical force. If the electrical force were equal to the distance to the nearest star, then the force of gravity would be 100 billion times smaller than a hydrogen atom. In what you might consider a strong electrical charge, such as a charge that literally will make your hair stand on end, the imbalance between positive and negative electrical charges is only about 1 in a billion billion.

difficult not to be struck by some of the surprisingly fortuitous accidents without which our existence would be impossible."[3]

Davies estimated in 1982 that Ω was, at this point in the history of the universe, between 0.01 and 9.0. Even so, when extrapolating back to the big bang, he noted that Ω had to be astonishingly close to 1, to within 1 part in 10^{60}.[4] (We now know that Ω is within 4 percent of 1, so the fine-tuning at the creation of the universe had to have been greater than what Davies calculated.)

There are other constants of physics that seem to be set just right; one list has 31.[5] When I asked my friend Peter Fisher, who is currently head of the Physics Department at MIT, which of all these coincidences seemed most amazing to him, he said what impressed him the most was how close Ω was set to exactly 1. One in 10^{60} is like one proton compared to the protons in a thousand Suns. Or imagine marbles, one-half inch in diameter, extending out in all directions 50 light-years from Earth. That ball, 600 trillion (600,000,000,000,000) miles in diameter, has about 10^{60} marbles. That's the precision of the gravitational constant alone.

As far as we know, the 31 or so fundamental constants of physics that have been set just right for life are unrelated. There is no reason to think any one of them can be calculated from any of the others. Yet somehow they are all just about perfectly right. The analogy is given of walking into a control room for the universe and finding that all the dials had been set precisely for life. You would not think it was a lucky accident. The most likely explanation would be that some intelligent being had adjusted the dials.

A number of distinguished scientists have converted from Atheism to Belief because of the fine-tuning of the universe. Fred Hoyle was astonished at how precisely nuclear resonance levels were set for the production of carbon and oxygen within stars. He concluded that "a superintellect has monkeyed with physics."

More generally, we might ask why exactly it is that we even have the right sort of a universe, with the right forces and rules, for life to exist. Why, for example, do we have three directions of space? In a universe with two space dimensions (like an endless sheet of paper), the necessary connections of life could not be made, and in a universe with four or more space dimensions (don't even try to imagine this!) gravity and electromagnetism would not follow the inverse square law (the force is weaker in proportion to the square of the distance), and planets and electrons would not have stable orbits.

Why does the force of gravity exist so that matter will gather into clumps? Why is there an electrical force to power the reactions and machines in our body? Why are there nuclear forces so that atoms can form? And so on. What causes it all to be so amazingly perfect? To me, the discovery of the fine-tuning of the universe is the second wonder of modern science. There is certainly disagreement as to what this means, but there is no real disagreement with the existence of fine-tuning, even among militant Atheists. Both the laws and the constants of physics give the appearance of having been designed for the existence of life. Physicist Freeman Dyson wrote: "The more I examine the universe and the details of its architecture, the more evidence I find that the universe in some sense must have known we were coming."[6]

Even noted Atheist Stephen Hawking, who does not think science requires belief in God, is impressed:

> The laws of science, as we know them at present, contain many fundamental numbers, like the size of the electric charge of the electron and the ratio of the masses of the proton and the electron. . . . The remarkable fact is that the values of these numbers seem to have been very finely adjusted to make possible the development of life.[7]

The fine-tuning of the constants of physics is now accepted as scientific fact. According to astrophysicist Michael Turner, "[T]he precision is as if one could throw a dart across the entire universe and hit a bulls-eye one millimeter in diameter on the other side."[8]

Besides the constants of physics, other items in our universe seem just about right. One is the chemical compound we call water. Water is believed to be a common chemical compound (although, as we will see in chapter 13, liquid water may be unusual). For example, in July 2011 researchers claimed to have discovered a cloud of water vapor 12 billion light-years away, with "140 trillion times more water than all of Earth's oceans combined," and stated the "discovery shows that water has been prevalent in the universe for nearly its entire existence."[9] Water makes up only .02 percent of Earth by weight, but it covers 71 percent of Earth's surface. (Keep in mind that although the average depth of the world's oceans is 12,000 feet, it's about 4,000 miles down to the center.)

To me, the special properties of water suggest design. According to Michael Denton, "If the properties of water were not almost precisely what they are, carbon-based life would in all probability be impossible."[10] Life is a chemical system that operates in water. Our bodies are mostly water. Water is pretty much the universal solvent, and it is now thought to be the only solvent in which life can arise.[11]

A molecule of water has two hydrogen atoms and one oxygen atom. It is not symmetric; the side with the two hydrogen atoms has a slight positive electric charge, and the side away from the hydrogen atoms has a slight negative electric charge. For this reason, water is cohesive, it sticks together, it has surface tension. For this reason, water breaks apart and dissolves compounds, like salt, that are held together by loose electrons. Molecules that are held together by shared electrons (covalent bonds) are not affected. Because oil (an organic compound) and water don't mix, the proteins in our bodies don't dissolve.

Another amazing feature of water is that ice floats. As liquid water cools, it is most dense at about 39 degrees Fahrenheit. Sometimes when you're out swimming, you can tell that the water just a few feet below you is cooler. But when water freezes, it expands about 9 percent, and ice floats. Were it otherwise, rivers and oceans would freeze from the bottom up, and many scientists think life on Earth would not have survived. Some scientists believe there were periods, before 650 million years ago, when the entire surface of Earth froze over.[12]

Water also has a fairly wide range of temperatures when it is in liquid form. The range between its freezing and boiling points is 100 degrees Centigrade and 180 (212 minus 32) degrees Fahrenheit. By comparison, the range for methane is 21 degrees Centigrade. It takes a lot of energy compared to most compounds to heat water up, so water holds heat well and keeps our bodies warm.

I want to conclude this chapter with one more example of exactly how fine-tuned our universe appears to be. It comes from Roger Penrose, a respected English physicist. He calculates that at the moment of the big bang, the universe was highly ordered. One result of the very high degree of order is the second law of thermodynamics—the amount of disorder is always increasing. Penrose states that the phase state had to be precisely calibrated, to something like 1 part in a number with 10^{123} zeros, to create a universe as special as ours.[13] You have to be careful how you calculate disorder—for example, heat is a kind of disorder—but it can be done. For the universe to have evolved to look as it does now, the initial conditions had to be set in a very special way.

This really is an unimaginable fine-tuning. As Penrose notes, one could not even write the number down, because the number of zeros is much greater than the number of particles in the visible universe.

This is such fine-tuning that the human mind can hardly begin to comprehend it. One part in 10^{27} is a marble one-half inch in diameter compared to the volume of Earth. There are about 10^{80} subatomic particles, such as a proton, neutron, or electron, in the entire visible universe.

Here's an example that may help you visualize how big Penrose's number really is, how incredibly unlikely it is to win a lottery where the odds are 1 in a number with 10^{123} zeros. We'll get there in four steps. The first step is you win a lottery where your odds are 1 in 10^{80}. This is like picking the exact lucky particle out of the entire visible universe, winning a Powerball lottery nine or ten times in a row, or picking the ace of spades out of a deck of cards forty-six times in a row.

For the second step, you have to win this same impossible lottery, picking the exactly correct "lucky" particle out of the entire visible universe, every single second for the entire age of the universe: 14 billion years. Every single second for 14 billion years you are lucky enough to pick the particular particle in the universe that won you this impossible lottery. That amount of luck gets you to step two.

In his celebrated volumes, historian Will Durant opens with the following image of time:

> High up in the north, in the land called Svithjod, there stands a rock. It is a hundred miles high and a hundred miles wide. Once every thousand years a little bird comes to this rock to sharpen its beak.
> When the rock has thus been worn away, then a single day of eternity will have gone by.

Let's play with this image. Suppose our little bird, when "sharpening" its beak, dislodges only one atom from this immense rock.* This is such a small loss that, even after a trillion (1,000,000,000,000) visits taking

---

* I imagine here that the rock is made of solid hydrogen, the most common element in the universe. A hydrogen atom has a diameter of 10^{-10} meters, so I assume it takes 10^{30} visits for the bird to wear away one solid meter.

atoms from the same spot, the change would probably not be visible to the naked eye. Suppose our little bird doesn't scratch his beak every thousand years, but instead comes only once every 14 billion years. And suppose that, as these unimaginable eons slip by, every single second you are "lucky" enough to pick the right particle in the entire visible universe. Every second for 14 billion years you pick the exact right particle in the universe—perhaps one second an electron passing through the Andromeda Galaxy, perhaps another second a proton being swallowed by a black hole 10 billion light-years away, and so on—and all of this luck corresponds to just one atom coming off of this massive rock every 14 billion years. Let me say that again: to take a single atom off of this rock, you have to be "lucky" enough to pick the right particle in the entire visible universe every second for 14 billion years, you have to have the "luck" of step two. To get from step two to step three, you have to repeat the "luck" of step two without fail as many times as there are atoms in this massive rock. By the time the little bird has worn away the entire rock, we are approaching the fine-tuning of the big bang, fine-tuning to 1 part in a number with 10^{123} zeros.

We need just one more adjustment. For step four, instead of the rock being a cube 100 miles high, 100 miles wide, and 100 miles deep, imagine a rock that is three hundred million (300,000,000) light-years high, three hundred million light-years wide, and three hundred million light-years deep. That's about two billion trillion miles of solid rock, or 3 times 10^{24} meters of solid rock, in each direction. And that little bird comes along once every 14 billion years to scratch one atom off of it. Meanwhile, until the entire rock has been worn away, somehow you are "lucky" enough every second to pick exactly the right particle in the entire visible universe, time after time after time after time. That gives you a sense of the precision, the "luck," the fine-tuning, that Roger Penrose calculates took place at the big bang when our universe was created.

What would the odds have to be for you to believe in God?

CHAPTER 9

Problems with the Multiverse

Is there more than one universe?

Only two things are infinite, the universe and human stupidity, and I'm not sure about the universe.

Attributed to ALBERT EINSTEIN

To MOST PEOPLE, the word *universe* means "everything"—the totality of existence, the complete contents of intergalactic space. Everything we will ever be able to observe or measure is in our universe; it is hard to see how we could ever observe or measure something not in our universe.

Yet many scientists believe our universe is an insignificant part of a much grander scheme, a collection of universes they usually call the "multiverse." These scientists believe existing universes can somehow spin off new universes. For example, they suggest new universes may be created inside "black holes"—objects so massive that nothing, not even light, can escape. Perhaps, but since there is no way to test whether it's true, it's clearly just a belief. These scientists believe that in another part of the multiverse, in a different "universe," the rules could be different. There could be different forces and constants of physics, perhaps even different dimensions.

The multiverse has enormous, intuitive appeal to Atheists. You don't have to be an Atheist to believe in the multiverse; there are people who believe in both God and the multiverse, but if you are an Atheist, you're likely loving the multiverse concept. If you believe in the multiverse, then our universe is not unique, and perhaps not so special.

Atheists typically blend two further beliefs into the multiverse concept. First, they believe the multiverse contains an infinite number of universes. This belief allows them to sidestep the philosophic problem of the first cause. If there were only a finite number of universes, however large, then at least one of those universes would have to be an original, a great-granddaddy universe not created by any other universe, and Atheists would have to confront the unanswerable (within the Atheist worldview) question of what caused that great-granddaddy universe to exist, what brought it into being. Atheists don't like that question, so they generally sidestep it by believing that the multiverse contains an infinite number of universes, universes without beginning and without end.

The second belief Atheists add to the multiverse concept is that the constants of physics, and even the laws of physics, change from one universe to another. Atheists believe this universe-creating mechanism somehow has the ability to spew out new laws of physics with different constants. This belief, coupled with the belief in an infinite number of universes, allows Atheists to sidestep the scientific fact that our universe is fine-tuned for the existence of life. If you believe in this kind of multiverse, then fine-tuning is not surprising—in an infinite number of universes some would be expected to have the constants of physics set just right to allow life to exist, and since we are life in our universe the laws and constants of physics are set just right.

When you stop to think about it, this universe-creating mechanism would have to be incredibly powerful and versatile. In our universe, the number of spatial dimensions (three) is just right, a point that Martin Rees makes in his book *Just Six Numbers*.[1] In a universe with two space dimensions (like an infinite piece of paper), the necessary connections of life could not be made, and in a universe with four or more space dimensions (this is hard for anyone to imagine), gravity and electromagnetism would not follow the inverse square law (the force is weaker in proportion to the square of the distance), and planets and electrons would not have stable orbits. One dimension of time seems just fine. Our four fundamental forces seem to work well. The strong and weak nuclear forces bond the nuclei of atoms together; the electrical force powers all chemical reactions, including how we walk and talk and think; and the gravitational force creates galaxies, stars, and planets. The proton, the neutron, and the electron—the three basic particles of ordinary matter—seem to work amazingly well together to create all the elements and the structure in our universe.

If we are to believe that none of these features was designed, then we need to believe the multiverse contains every possible variation. The multiverse must create new laws of physics, based on different numbers of dimensions, as well as new and varied concepts of space and time, different fundamental particles, and so on, probably in more ways than we can currently express or know, because we are still learning how our universe works.

Also, it would seem, at least to me and to many others, that anything capable of creating universes with different features and constants would have to be more complicated than the universes it creates. A bread machine is more complicated, and requires more technology and design, than a loaf of bread. An automobile factory is more complicated, and requires more technology and design, than the automobiles that roll off the production line.

So to summarize the discussion so far, this Atheist multiverse includes a number of fundamental *beliefs*:

- Our universe is one of an infinite number of universes.
- There is some type of universe-creating mechanism that spews out new universes with totally different and alien dimensions, laws, features, and constants of physics.
- Somehow, for no special reason, all of this just exists.

There is a principle of science called "Occam's razor," which loosely stated suggests that explanations with a smaller number of unproven assumptions are more likely to be correct. To me, belief in God is simpler than the Atheist belief of an infinite multiverse with changing laws, dimensions, constants, and features of physics. The multiverse has been called a "flagrant violation" of Occam's razor. I agree with that statement, but that's just me, and I'm sure others see it differently. It also seems clear that, given the enormous chasm between the worldviews of Belief and Scientism, perceptions of simplicity are not likely to be a deciding factor.

What is most strange about the multiverse concept is the refusal by many scientists to admit it is only a belief. There is no evidence that other universes exist or that the laws and constants of physics can change—absolutely none. Yet hundreds of peer-reviewed articles on the multiverse have been published in scientific books and journals. If you count articles "published" simply by posting on Internet sites, there are thousands of scientific articles on the multiverse.

What strikes me as plainly wrong is that these articles rarely admit the multiverse is an unsupported belief. The multiverse is science fiction, not science fact. Yet magazines run headline articles about the multiverse. I think many people see the headlines and assume that if it's discussed in science magazines, it must be real. Even if you read the articles closely, they rarely admit there is no scientific evidence that the multiverse exists.

Stephen Hawking tells of a lady who, after a lecture about the universe, stated she believed Earth was supported by a giant tortoise. When asked what the tortoise was supported by, she responded: "Very clever young man, but it's turtles all the way down." Hawking perhaps intended to suggest that there are a lot of strange theories about the universe. But I would turn it against him and other multiverse believers: what supports the multiverse? To me, the infinite multiverse is turtles all the way down, a preposterous science fiction concoction.

I also think it is wrong, and a clear error in logic, to suggest, as many Atheists have done, that these theories about and beliefs in multiple universes somehow confirm or even imply that God does not exist. It is interesting to note that, while any suggestion of design in the universe is immediately attacked as unscientific, these Scientism, Anti-Faith articles that profess belief in multiple universes are widely accepted and sail through peer review. There is a clear bias against belief in God. If one wants to believe there are an infinite number of other universes with different laws, dimensions, constants, and features of physics, and all of this just exists for no reason, then to be intellectually honest, one should admit that belief in God is at least equally reasonable. The accepted scientific fact of creation implies that something exists outside of our universe, outside of our reality. There is no reason to think we will ever have scientific proof that other universes exist. There is ample room for wonder.

Believers in the multiverse generally ignore a subtle but serious mathematical problem. The problem arises because they believe the number of other universes is infinite.

Infinity is strange. It exists only as an idea, as a mathematical concept. As far as we know for sure, nothing in this universe is infinite. Infinity is not some cute but shy number always hiding over the next hill. Infinity is not just a number that is always a little out of reach. Infinity is a monstrously alien concept. Carl Friedrich Gauss (1777–1855), perhaps the greatest mathematician to ever live, said that one should not attempt to look directly at infinity.

Suppose there were a number measurement scale with a dial. Suppose at one end of the scale is the number 1, and at the other end is infinity. So when we "measure" the number 2, it looks like the illustration below.

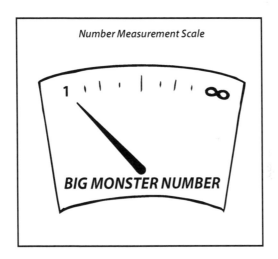

The dial points to the same place as 1. It has not moved in any way closer to infinity. When you go from 1 to 2, you don't get any closer to infinity.

Now let's "measure" the biggest number you can possibly think of. It has to be a number that can be expressed in a finite number of steps (you can't say "add one to itself an infinite number of times"). We got a sense earlier of how big a number with 10^{123} zeros is. Suppose we start with one trillion—1,000,000,000,000—and then take 10 to that number (create a new number with 1 trillion zeros). Then take that number and make it the exponent of a new power of 10 (create a new number that has $10^{1,000,000,000,000}$ zeros). We continue this process, where we keep taking 10 to a power equal to the prior number (we use the prior number as the exponent for the next), and we do that 1 trillion times. By the third step, much less the trillionth step, we have a number hugely bigger than a number with 10^{123} zeros. Suppose the number we get after doing this 1 trillion times is my "big number," and suppose you and every person on Earth creates their own "big number." Now imagine we multiply them all together, the "big numbers" of every person on Earth. Call this our "Big Monster Number." Let's put it on our number measurement scale.

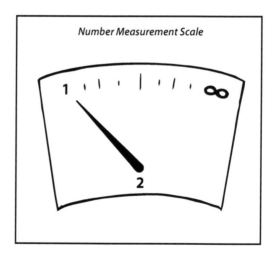

Whoa! The Big Monster Number has not budged the dial! It is no closer to infinity than the number 2. You cannot get partway to infinity. However far you go, you are still no closer to infinity than when you started. The gap between finite numbers and infinity cannot be crossed. That is the essence of why infinity is such a strange concept. In all of our human experience, when we move toward something, we get at least a bit closer. Infinity doesn't work that way.

When you realize how alien and monstrous the concept of infinity actually is, I think it becomes very hard to "believe" there really are an infinite number of other universes. Infinity can break itself up into an infinite number of groups, each of which is also infinite and every bit as large as the original infinity.* If something can ever possibly happen with any chance whatsoever, no matter how small, then in an infinite number of universes it has happened infinitely often. If there are an infinite number of universes, then, regardless of how small the probability may be that one of them is a universe with exactly the same physical laws as ours, there are an infinite number of universes exactly like ours. But, as they like to say on TV, "wait, there's more." If there are an infinite number of universes exactly like ours, then, regardless of how small the probability may be that exactly all the atoms in your body are put together exactly right to make you, there are or have been an infinite number of persons exactly like you. Can you really believe there have been an infinite number of universes just like our own, with persons just like you in every respect, with your exact name and history, who are doing or have done exactly what you are doing now? If you believe in an infinite number of universes, then you must. Infinity is a monster that cannot be tamed in the real world. However, in the world of ideas, the world of number, infinity can be fun. Appendix B illustrates why infinity is such a strange concept and how mathematicians use it.

This chapter began with a question: Is there more than one universe? Science can't answer that question, not now and probably not ever. So you have a choice of beliefs. Each belief has logical consequences.

If you choose to believe our universe is all there is, then something outside our universe caused our universe to come into being. That something, that first cause, could be God. It could be the God of the Bible. So this belief is consistent with Abrahamic faith. The fine-tuning of this universe, the proven fact that it is just about perfect for life, is also consistent with Abrahamic faith.

If you choose to believe there are other universes, but a finite number of other universes, then at least one of them was not created by another universe. At least one universe just is, just exists. What caused that universe to exist? What was its first cause? That something, that first cause,

---

* I know this sounds strange; see appendix B.

could again be God, and could again be the God of the Bible. So this belief is also consistent with Abrahamic faith, but clearly it is not how most people interpret the Bible, and here also the fine-tuning of our universe is perhaps less of a surprise.

If you choose to believe (1) that we are part of a multiverse that contains an infinite number of other universes, (2) that this multiverse has always existed, and (3) that this multiverse has the ability to create new universes with different laws, dimensions, constants, and features of physics, then you can sidestep the philosophic implications of a first cause and the fine-tuning of the universe that we happen to be in. But you must recognize the mathematical problem described above; there are an infinite number of persons exactly like you in every respect. And you are no closer to the first cause. What caused that unwieldy multiverse to exist?

Atheists say it just does. Perhaps, but that is a *belief.*

CHAPTER 10

The Origin of Life

How did life begin?

Nobody knows how it got started.

RICHARD DAWKINS

And God said, "Let the water teem with living creatures, and let birds fly above the earth across the vault of the sky."

GENESIS 1:20

RICHARD DAWKINS MAY BE the most vocal Atheist of our day. I credit him in his admission above for honesty. Science alone has no explanation for the origin of life. To me, the discovery of the complexity of life, and our inability to conceive of even a plausible pathway for its origin, is the third wonder of modern science, the third of seven in our count to God. Life is a miracle. In the creation of life—the most primitive, simplest form of life—we see the hand of a master designer.

At one point on my journey into the science of belief, I decided I needed to learn the basics of molecular biology. I ordered the standard college treatise, 1,462 oversized pages of single-spaced text, and started reading.* Fun, fun,

---

* That's the "fun" *Molecular Biology of the Cell,* 4th ed. (New York: Garland Science 2002). The "more fun" fifth edition (2008) has about 1,600 pages but provides the last few hundred on a DVD.

and more fun. OK, it was not what most people would consider fun, but to me it was riveting. It was all there, all the details about how life runs because of biological parts we call proteins, how life builds proteins using DNA, and how you need a lot of different proteins (perhaps 100,000 or more different types for a human being), and a lot of complicated machines built by putting those proteins together just right, to have life. Proteins are special molecules. At one point, the math nerd in me could not help but calculate, literally on the back of an envelope on an airplane, the fantastic improbability that a single functional protein was ever created by accident in the entire history of the universe. I was thunderstruck—it was an "Aha" moment. I remember staring at the calculations in disbelief—couldn't others do the math, and see what seemed obvious? It was a "no-brainer." At that moment, I knew modern science supported belief in God.

The fantastic improbability of the origin of life is shocking. The calculations are relatively simple, yet compelling; I'll show you how to do some of them in this and the next chapter. One might perhaps choose to believe that other evidence for the existence of God, such as the fine-tuning of the universe, can be explained by laws of physics we don't yet understand. But we know well the laws of physics and chemistry that govern the creation of molecules, and we can estimate the probability that certain specific and complex combinations were formed by accident.

This chapter and the next two chapters reveal the technology of life. This chapter focuses on the origin question—How did life begin? It is the Achilles heel of neo-Darwinian belief. It is the question that won't go away, the question Scientism can't answer.

We'll set the stage with ground rules. Then we'll follow my journey. We'll look at the Miller-Urey experiment, how in high school it hastened my path to Atheism, and why it is no longer considered good science. We'll look at Harvard's failed Origins of Life Initiative. We'll look at evidence I find fascinating if circumstantial—the Search for Extraterrestrial Intelligence (SETI) program and the search for life elsewhere in the universe. And we'll begin to calculate the fantastic improbability of life.

Ground Rules

As we enter the debate, we need to remember two rules. First, you have to apply the chemical and physical laws we know today. You can't say life arose because of some unknown law of chemistry or physics, or because the laws of chemistry and physics were wholly different billions of years ago. Scientists agree that the chemical and physical laws we observe today have not changed in any significant way since the creation of the universe. This is known in part from studies of natural low-intensity nuclear reactions that have taken place in our Earth for billions of years, and from light emitted billions of years ago by distant "quasars."* In fact, it is commonly believed that the natural laws we observe today also governed the very early universe less than one trillionth of a second (.000000000001 second) after the big bang.

The second rule is you can't use Darwin's theory of natural selection to explain the origin of life. Natural selection—sometimes called (slightly inaccurately) "survival of the fittest"—is the concept that traits which give an organism an increased ability to reproduce are more likely to be passed on to succeeding generations. To have natural selection, you must first have a means of inheriting traits across generations. Before life existed, there was only inorganic matter with its chemical and physical properties. There was no way to preserve and reproduce successful traits in succeeding generations.

Darwinian theory at most explains life's ability to adapt.[†] It cannot explain the origin of life. Inorganic matter does not "naturally select." There is, and was, no "most likely to succeed" form of inorganic matter. This seems obvious, but many otherwise reputable scientists refuse to accept it. They argue that the molecules most likely to eventually become life developed by natural selection. Hogwash. Inorganic molecules form, and change, according to the known laws of chemistry and physics. Inorganic molecules do not compete with each other for food; they do not pass their genes on to other inorganic molecules; they do not have a means for passing on successful traits.

---

* Recent observations of distant quasars have led some to conclude that there could have been very slight changes in the constants of nature, on the order of one part in a million, when the universe was very young, billions of years before life arose on Earth. See http://arxiv.org/abs/1202.6365. This is disputed by others. I find this suggestion fascinating. Either way, it strengthens the argument here, by confirming that there have been no major changes in the relevant laws or constants of physics for billions of years.

† Scientists have discovered that many mutations are "directed mutations," where internal programming in an organism causes it to adapt or mutate as necessary to survive. This is an exciting new area of science revealing design. See chapter 11, note 2, on page 231.

For example, one author talks of inorganic "catalysts" evolving so that "[t]hose that were best at promoting their own production (and inhibiting their own destruction) relative to other variants became more numerous. They too promoted the construction of variants of themselves, and so evolution continued."[1] This sounds plausible until you realize there is zero science behind it; it is just the author's naked belief that someday molecules will be found that violate the second law of thermodynamics and become ever more complex instead of degrading. As Michael Behe has pointed out, "There is no publication in the scientific literature in prestigious journals, specialty journals, or books that describes how molecular evolution of any real, complex, biochemical system either did occur or even might have occurred. There are assertions that such evolution occurred, but absolutely none are supported by pertinent experiments or calculations."[2]

Life requires fantastically complex molecules, and inorganic molecules do not become ever more complex over time. Complex molecules break down and degrade. The largest, most complex molecules ever found in meteorites or in outer space are trivial compared to the molecules found in all life on Earth. The largest hydrocarbon molecule found in space is anthracene, with 14 carbon atoms and 10 hydrogen atoms, and the largest molecule of any kind is 70-fullerene, with 70 carbon atoms.[3] The machine parts of life—proteins—typically contain thousands of atoms of multiple elements (carbon, hydrogen, oxygen, nitrogen, etc.) precisely arranged. Such molecules have never been found in outer space.

We must approach the origin of life with science. No magic tricks. Scientism claims life arose by accident. In this chapter we ask: what are the odds?

The Miller-Urey Experiment

I was taught in high school that scientists had shown life could have arisen on the ancient Earth without a designer. The basic concept was lightning struck some ancient pond billions of years ago, and life was created by accident. Totally by accident, no design, no purpose. We all know accidents happen. You fall and stub your toe; lightning interrupts the television show you were watching; lightning hits some primeval ooze and creates life. A simple accident.

What was not accidental was the effect this "scientific" knowledge had on me. It hastened my shift from God to Atheism. I believe this scientific misinformation has had a similar effect on millions struggling with their personal version of the great debate.

The experiment I was taught is called the Miller-Urey experiment. Stanley Miller was a graduate student at the University of Chicago. In 1953 he and his faculty advisor, Harold Urey, attempted to reproduce conditions on the surface of Earth billions of years ago. They mixed together chemicals in the proportions they believed were present in the early Earth. To this mixture, they added electrical charges, and they found that certain amino acids were produced. As we will see, amino acids are the building blocks of proteins, and proteins are the machine parts of life. This was reported as the first time an "organic" molecule had been produced from inorganic lab chemicals, and it caused quite a stir. The neo-Darwinian thought police, aided by the popular media, immediately seized upon this experiment as "proof" of their theories. The barrier between nonliving matter and life had been broken, they claimed. Without a great deal of critical analysis, this production of amino acids was hailed as sort of a missing link validating Darwin's theories.

It was and remains a serious blow in the minds of many against the side of Belief. Misplaced belief in the Miller-Urey experiment continues. A 2011 study of twenty-two high school textbooks found that nineteen discuss the Miller-Urey experiment as a possible explanation of the origin of life.[4] I have met many people who think life began simply because lightning struck some primeval pond billions of years ago. The Disney movie *Fantasia* contains such a scene, set to the music of Stravinsky's "Rite of Spring."

Despite this popular perception, the Miller-Urey experiment is no longer considered good science. We have known for decades that all life on Earth requires, among many other things:

1. Twenty different "amino acids." These are special groups of carbon, hydrogen, oxygen, and nitrogen atoms. There are about 500 different types of amino acids; every living creature uses exactly the same 20 amino acids to build proteins.
2. A mechanism for linking amino acids together into long chains, to build the molecules we call "proteins."
3. Special, hard to produce chemical groups called "nucleotides," which also include phosphate atoms, to store the code of life.

4. The information to assemble all of these groups of atoms into the right three-dimensional shape and the right order, so that they simultaneously become both the code for life and the machinery to process the code of life and do the work of life.

The Miller-Urey experiment fell far short of demonstrating how any of these could have occurred by accident or mere chance.

The hypothesis that life formed by pure chance in some primeval pond is contrary to modern science. For starters:

1. Wrong atmosphere: Consensus views of Earth's early atmosphere have changed. First, Miller assumed no oxygen and ample free hydrogen. Both of these assumptions are now questioned; more recent analysis suggests some free oxygen was present in the atmosphere of the early Earth. Although oxygen is essential for life today, it is thought life could not have formed in the presence of free oxygen, because oxygen reacts quickly with many organic compounds and would have destabilized early molecules. Second, Miller also included chemicals rich in hydrogen, such as methane and ammonia. There is little or no current evidence that such chemicals existed in substantial quantities on the surface of the primeval Earth. According to *Science* magazine in 1995, "[T]he early atmosphere looked nothing like the Miller-Urey simulation."[5]

2. No "primordial" soup: While all agree the early Earth was a hot, chemically active place, it is not clear how the building blocks of life, if they were accidentally created, would not have rapidly degraded or been destroyed by ultraviolet radiation and other factors. In addition, some of the chemicals produced by the Miller-Urey experiment were not conducive to life, such as hydrogen cyanide, which is extremely poisonous, and formaldehyde, which is highly toxic. The primary "organic" product of the Miller-Urey experiment was tar. The primordial soup theory "doesn't hold water" and is "past its expiration date."[6]

3. No way to link amino acids: Amino acids do not automatically link together. Amino acids are like plastic blocks that snap together; it takes energy. To quote the National Academy of Sciences, "Two amino acids do not spontaneously join in water. Rather, the opposite reaction is thermodynamically favored."[7]

4. No information: Where did the information to build life come from? We will focus on this below.

No serious solution to any of these problems has been proposed. One mathematician has stated that understanding the origin of life "cannot be bridged within the current conception of biology."[8] Even Fred Hoyle, the once Atheist who originally mocked the theory of creation of the universe and sarcastically called it the "big bang," agrees: "In short there is not a shred of objective evidence to support the hypothesis that life began in an organic soup here on the Earth."[9]

The concept of the Miller-Urey experiment—that life could have formed somehow from purely natural and random events—may far outlive belief in the validity of the experiment itself. Even if Miller and Urey got it all wrong, the concept that inorganic chemicals accidentally combined to create the first living organism has, for many, a powerful appeal. Because we don't know all the ways chemicals and compounds might have been concentrated on the early Earth, it is reasonable, at least in my view, to cut the Miller-Urey experiment a little slack. For purposes of argument, let's consider the possibility that amino acids could have been created by accident or random events, and perhaps in great numbers. (But, again, there is no scientific basis for concluding that all of the amino acids required for life could have been produced by accident or linked together by accident.)

The mere existence of amino acids does not create life. "A mixture of simple chemicals, even one enriched in a few amino acids, no more resembles a bacterium than a small pile of real and nonsense words, each written on an individual piece of paper, resembles the complete works of Shakespeare."[10] To create life, you need amino acids plus a lot of information, information on how to assemble the amino acids and other complex molecules to form life. How could hundreds of thousands of amino acids have arranged themselves, with other necessary complex molecules (including some that are extremely unlikely to arise by accident), to create life? And how likely is it that these very complex molecules, many of which work together with amazing precision to read a chemical code and manufacture themselves and other molecules, accidentally formed at the same time and in the same place as the precise chemical code that they are able to read, and that contains the precise instructions to build these exact molecules and other necessary proteins?

Both Miller and Urey later admitted the mere existence of amino acids does not yield life. Harold Urey came to this conclusion in 1962:

> [A]ll of us who study the origin of life find that the more we look into it, the more we feel it is too complex to have evolved anywhere. We all believe as an article of faith that life evolved from dead matter on this planet. It is just that its complexity is so great, it is hard for us to imagine that it did.[11]

So did Stanley Miller, in a paper he coauthored in 2007:

> The origin of life remains one of the humankind's last great unanswered questions, as well as one of the most experimentally challenging research areas.... Despite recent progress in the field, a single definitive description of the events leading up to the origin of life on Earth some 3.5 billion years ago remains elusive.[12]

Harvard's Origins of Life Initiative

Of course, not everyone in the predominantly Atheist, scientific elite has admitted defeat. Since 2006, Harvard University has sponsored an "Origins of Life Initiative."[13] Harvard would surely disagree, but to me, and to many scientists, it looks like they are dead in the water. As of mid-2012, before the initiative took down its list of research papers, its website had included no new biology papers in three years.

In March 2009 the Origins of Life Initiative brought together a number of distinguished scientists to discuss how life could have begun. Here's how it went:

> It may be difficult to believe, but there was a common theme to this seeming cacophony of scientific expertise and discovery. The theme was, "We just don't know." No one knows how life began. . . . Underneath it all, it was refreshing to hear a bunch of really smart folks say "we don't know." It was humbling and put things in a grandiose perspective.[14]

What's amazing is not that we don't know precisely how life began. What's amazing is that our most brilliant and motivated scientists, drooling over the possibility of eternal fame and a Nobel Prize for solving just a piece of the riddle, can't come up with a mildly "plausible" scenario. They really are dead in the water. If you don't believe me, here's a 2011 status report from Eugene Koonin, a senior investigator at the National

Center for Biotechnology Information, National Library of Medicine, National Institutes of Health, and a recognized expert in the field of evolutionary and computational biology:

> Despite many interesting results to its credit, when judged by the straightforward criterion of reaching (or even approaching) the ultimate goal, the origin of life field is a failure—we still do not have even a plausible coherent model, let alone a validated scenario, for the emergence of life on Earth. Certainly, this is due not to a lack of experimental and theoretical effort, but to the extraordinary intrinsic difficulty and complexity of the problem. A succession of exceedingly unlikely steps is essential for the origin of life, from the synthesis and accumulation of nucleotides to the origin of translation; through the multiplication of probabilities, these make the final outcome seem almost like a miracle.[15]

Hmmm. So it seems like a miracle and it looks like a miracle. Perhaps it was a miracle. Perhaps it was an act of God.

The Harvard scientists published suggestions a few years ago that certain fatty acids could have served as membranes for the first living cells. They cannot explain how complex proteins, and the required DNA or other coding to build those same proteins, could have simultaneously formed inside any membrane. Perhaps frustrated by this inability to penetrate the molecular biology obstacles to the origin of life, the Harvard Origins of Life Initiative has changed course toward the search for Earthlike planets. The initiative redesigned its website to focus on astronomy and away from the intractable problems of the origin of life.

Before we begin to estimate the fantastic odds that life was created by accident, even assuming all the right chemicals were present, let us, like Harvard, turn to a related, although ultimately inconclusive, question. The universe is a big place. If life arose by accident on Earth, and if there are billions of Earthlike planets out there, surely it arose by accident in other places. So where is it?

Extraterrestrial Life?

If we step back from chemistry for a moment, and turn to astronomy and the search for life elsewhere in the universe, we find other evidence that perhaps the creation of life was special, perhaps an act of God. Let's start with our cosmic neighbor Mars.

For many years, Mars was thought to be the most likely place for life elsewhere in the solar system. Mars is somewhat similar in size and composition to Earth, and the next planet out from the Sun. As early as 1887, some astronomers thought they saw lines on Mars that were canals created by an intelligent civilization; the astronomer Percival Lowell drew sketches of intricate canal systems. Although many astronomers were skeptical and suggested (correctly) the canals were an optical illusion, the idea captured public attention. On August 27, 1911, the *New York Times Sunday Magazine* ran the following headline:

MARTIANS BUILD TWO IMMENSE CANALS IN TWO YEARS
Vast Engineering Works Accomplished in an Incredibly
Short Time by Our Planetary Neighbors*

This was published just over a hundred years ago, and after Einstein developed his theory of special relativity. You can't always believe what you read in the paper.

Mars has been the inspiration for delightful, and sometimes terrifying, stories of alien life. One of the most famous is the 1938 radio broadcast of H. G. Wells's *War of the Worlds*. It is a story of an invasion of Earth by fierce aliens from Mars. It was aired on Halloween, without commercials, in the realistic form of an interview about life on Mars interrupted by news bulletins, and there was widespread panic because millions did not realize it was fiction. The story was loosely adapted into a science fiction movie directed by Stephen Spielberg and starring Tom Cruise, which was the fourth most successful film at the box office in 2005.

As a child, I remember wondering if canals on Mars were real. The idea of canals on Mars did not die until 1965, when NASA's *Mariner 4* took detailed pictures that showed no evidence of canals or life—only rocky desolation. Yet microscopic life may be present on Mars, and that exciting possibility has contributed to the launch of Mars rovers. These are robots, controlled from Earth, with the ability to move about on the surface of Mars, take pictures, and perform sophisticated tests on Martian soil. Some of these tests are designed to detect life.

---

* In the article, Percival Lowell is quoted as saying that "[t]he whole thing is wonderfully clear-cut." He suggested dying Martians were building canals to reach the water ice in the Martian poles. See http://astroengine.com/2009/04/24/in-1911-martians-were-building-canals/.

August 5, 2012, witnessed the landing on Mars of NASA's newest and most sophisticated Mars rover: Curiosity. Curiosity is also looking for signs of life. It has discovered what scientists believe was an ancient stream, where running water once existed, and made an amazing discovery— conditions on Mars were once suited to life. "We have found a habitable environment that is so benign and supportive of life that probably if this water had been around and you had been on the planet, you would have been able to drink it,"[16] says one NASA scientist. What you don't read in the paper is the obvious conclusion to be drawn if Mars has no life: that the absence of life on Mars, despite once favorable conditions, suggests the creation of life on Earth was special.

But other evidence suggests there *is* life on Mars. Some believe NASA detected life on Mars in 1976. Here's how physicist Rob Sheldon explained the findings to me:

> It was called the labeled release experiment. It was developed over the past 50 years as a way to test water downstream from a sewage treatment plant to see if it was clean enough. It consisted of putting some sugar and test water in a flask, incubating and looking for bubbles—as the sugar was turned into carbon dioxide by the bacteria. Gil Levin suggested replacing the sugar with radioactive C^{214} labeled sugar. Then as the bubbles came off, a Geiger counter sampling the air above the sample would begin to chatter. This was not only quantitative, but very sensitive, because the CO_2 rate kept increasing as the bacteria grow, giving a very specific biologically produced growth curve. In 24 hours it could sense 2–3 bacteria per milliliter of water. In terms of mass, that is about 1 part in 10^{11}.
>
> By way of comparison, the mass spectrometer that Viking Lander used in 1976 had a ppm or 1 part in 10^6 or maybe in best case, ppb or 1 part in 10^9. So Gil's experiment was selected for both Viking landers, and performed perfectly. 2 samples were tested at 2 different sites, and a control was baked at about 150C for several hours. A total of 8 runs were made, the 4 test cases all showed the growth curves observed on Earth, the 4 controls were all flat-lines. And there was a sample that he let sit in the hopper drying out for 3 months, and when it was tested, it was also flat-lined. You can read all of his papers on his website, or go to the NASA Viking lander website and read up on "Labeled Release Experiment."[17]

I find this exciting. To me, life on Mars would in no way diminish the wonder of the origin of life. It is possible, some might say likely, that microscopic life has traveled between planets. Some speculate that life could have traveled within the solar system inside meteorites—rocks—that are blasted off the surface of one planet and ultimately captured by another.* There are meteorites on Earth that scientists believe came from Mars, and it seems possible that Mars and other planets in our solar system have captured meteorites from Earth, such as pieces of Earth blasted into space 65 million years ago in the meteor impact that may have ended the reign of the dinosaurs. We have learned that some types of microorganisms can survive in extremely harsh environments. If life is ever found elsewhere in our solar system, a key question will be whether it has the same DNA/protein design (see chapter 11) as life on Earth, and thus perhaps once traveled between planets.

Life on Mars would be a blow to our terrestrial chauvinism. The forces of Scientism would surely claim it demonstrates life is not special and that it could have arisen without a designer. But that is a belief without support in true science. There is no remotely plausible theory, under any conditions, for the origin of life by pure chance.

The idea of advanced life beyond Earth but within the solar system now seems fanciful. Its other planets and bodies are even poorer candidates. Some have suggested that life may exist on the moons of Jupiter or Saturn, particularly Europa, the smallest of the four moons of Jupiter discovered by Galileo, which appears to have an icy crust with the possibility of warm, liquid water below. Perhaps, but all of those moons suffer from huge variations in heat and cold that could destroy or damage life.

The search for life outside our solar system has been a total failure. In 1966 Carl Sagan stated there could be as many as 1 million advanced civilizations in our galaxy, although he reportedly backed off that number later. But if they're "out there," we have been unable to find them. The search for extraterrestrial life began over fifty years ago. The SETI Institute has methodically searched the heavens for extraterrestrial intelligence for over twenty-five years and found nothing. There is no evidence of intelligent life "out there."

---

* This is sometimes called "panspermia." For example, in August 2013 it was suggested that life may have started on Mars and then traveled to Earth, because of chemical properties of Martian rocks. See Simon Redfern, "Earth Life 'May Have Come from Mars,'" BBC News, http://www.bbc.co.uk/news/science-environment-23872765 (accessed August 30, 2013). This theory does not address the origin of the information necessary to create life. Its author admits that "[n]othing I am saying should be interpreted as the problem of the origin of life has been solved."

Our galaxy is 12 billion years old. Some estimate that a civilization as advanced as ours, or more advanced, would colonize the entire galaxy in 5 to 50 million years. Even if they did not want to "leave home," they could use self-replicating space probes—sophisticated machines that can explore and reproduce themselves on alien worlds. Such "Von Neumann probes" would allow an advanced civilization to explore the entire galaxy without leaving home, perhaps in less than a million years. We have found no aliens, we have found no probes, we have found no signals.

So where are they? One or 5 or even 50 million years is a blip compared to the 12-billion-year age of our galaxy (just 1 percent of 12 billion years is 120 million years). As we will see in chapter 12, our Earth is special. Surely an advanced civilization, or its probes, would have reached Earth millions if not hundreds of millions of years ago. An extremely advanced civilization might be able to alter the position or color of stars. Our telescopes reveal no evidence of that.

We have found absolutely no evidence that intelligent life exists anywhere else in the universe. Of course, this evidence is inconclusive; it does not prove that life or even intelligent life does not exist "out there." Perhaps there are a vast multitude of advanced civilizations that, for whatever reason, have no desire to make their presence known. Perhaps we are being "quarantined," left alone to manage our own fate. To me, our inability to find signs of life elsewhere in the universe is significant. It suggests the creation of life was a special event.[18]

Or you could go with the theory of my comic strip heroes Calvin and Hobbs—"The surest sign that intelligent life exists elsewhere in the universe is that it has never tried to contact us."

Let's go back to biology, and take a closer look at what we call life.

The Wonder of Life

It is hard, perhaps very hard, to appreciate how miraculous life is. We are life; we are surrounded by life. There is life within us, in the form of bacteria and viruses.* There is life in the air, and life in the sea. We

---

* Our bodies have an estimated 100 trillion bacterial cells, with a total weight on average of perhaps three pounds, that perform many essential functions to keep us alive. This book is not able to describe all the wonder and complexity of life, but our dependence on bacteria is yet another example of that wonder and complexity.

have found life in rocks, life around hot volcanic vents miles below sea level, and life in the arctic. To residents of planet Earth, life is everywhere; it is all around us.

Yet when you stop to ponder exactly what life is and how it works, you begin to realize how precious life is. From the perspective of engineering alone, life is an amazing chemical machine. It is a chemical machine that possesses the ability to reproduce—to make copies of itself, to make other machines of the same type. That is astounding. As advanced as our science and technology are, we human beings are not even close to building a different kind of chemical machine that can reproduce without assistance. Sure, we can "tweak" existing forms of life to enhance crop yields or obtain other benefits, but that is just a relatively slight modification to an existing form of life—an existing chemical machine. We do not have the ability to create, from scratch, a new design for a chemical machine that can reproduce. We are not even close to having that ability.

We take life for granted because it is everywhere. Our planet is overrun by biological machines. There are at least 10 million different types (species) of machines; some estimate that tens of millions of other types (species) have not yet been discovered. A giant sequoia is about 10^{27} times the size of a virus; that's like comparing Earth to a toy marble one-half inch in diameter.* All of these machines—from the smallest to the largest—are incredibly complex. There are so, so many incredible systems. Coordinated systems allow blue whales to dive thousands of feet below sea level without being crushed and sing complex songs that travel across oceans. Other systems allow bees to do a dance that tells other bees where to find the best sources of pollen. There are systems for hiding, systems for fighting, systems for reproducing, systems for getting food, systems for communicating, and so on. And perhaps the most amazing system in all existence—the human brain.

I compare life to a chemical machine, but all life, even primitive life, is so much more. To say life is a collection of chemical machines is like saying the works of Shakespeare are a collection of words, or that my

---

* The world's second-largest tree is believed to be the giant sequoia the President, with a height of 247 feet and a total volume of 54,000 cubic feet. "The World's Largest Trees," *National Geographic* (December 2012).

favorite piece of music, Rachmaninoff's *Piano Concerto No. 2,* is a collection of musical notes.*

Where did all this life come from? How did it come to be? Does life arise automatically from matter, as was once believed?[19] The theory of spontaneous generation of life predates Aristotle, and it was believed by many into the 1800s before disproven by scientists such as Louis Pasteur and John Tyndall.

Today, there are only two choices for the origin of life. Did life first arise by chance, an accidental consequence of random chemical events? Or, strange as it may sound, could life have been designed and created by intelligence? Accident or design? That is again the question.

The Odds of Life

No matter what the conditions were on the early Earth, or anywhere else, the basic problem with neo-Darwinism—the prevailing paradigm that life was created by accident—is one of information. Even the simplest form of life contains an enormous amount of information, far too much information to reasonably conclude it all happened by chance.

Let us overlook the proven flaws in the Miller-Urey experiment; assume for the moment the early Earth was somehow teeming with trillions of just the perfect twenty amino acids to form life. Let us also overlook the difficulties of putting all the right molecules together in exactly the right place at the right time. Let us overlook the issue of whether a permeable skin or membrane could have existed that somehow contained the first form of life.

To have life, you must have both the proper biological machine parts— proteins—and the exact code to build those parts. If you just have the machine parts but not the code to build more, you have something that

---

* Here's biologist Ann Gauger: "[W]e're accustomed to talking and thinking about the cell as made up of machines (hardware), with DNA as the software program that somehow determines the hardware. This is an advance over imagining the cell as a few simple chemical reactions. But it's still radically inadequate, if not obsolete, when trying to capture the reality of what we're discovering in the biological world. We're in search of more adequate conceptual categories. And the outcome will make our current descriptions look utterly inadequate. What we want to do is to catch up to the evidence, and get beyond our own, quite limited ways of speaking of these realities." http://www.evolutionnews.org/2011/06/life_purpose_mind_where_the_ma046991. html.

cannot reproduce, and you do not have life. If you just have the code but not the machine parts to read the code and build more machine parts, you again do not have life. The origin of life has been called a "chicken and egg" problem. To have life, you need to start with both.

One candidate for the first living creature is blue-green algae. It can both use the Sun's energy (photosynthesis) and convert nitrogen into organic compounds (called nitrogen fixation), and thus can enter a sterile and bionutrient-free environment (no sugars, starches, amino acids, etc.) and start growing (assuming, of course, that water, carbon dioxide, and the right minerals are present). It is also the most abundant life-form on Earth. There are various strains of blue-green algae, with an average of about 2,000 genes (including about 1,100 common to all strains)[20] and a relatively small "genome" of 1.6 million letters of DNA.[21] Life, even so-called simple life, is vastly more complex than we generally perceive.

Each of the 2,000 or so genes in blue-green algae contains the code for building a specific protein. As we will see in the next chapter, all life builds proteins by linking together amino acids. Twenty different types of amino acids are used to build proteins. They are linked together, end to end. "Simple" proteins in the most basic forms of life typically have about 300 amino acids linked end to end, and then folded exactly right into a three-dimensional shape, a working biological machine part, a functional protein.

We can calculate the odds that a specific functional protein containing 300 amino acids—a single biological machine part—was created by accident. At each link in the 300-unit chain, there are 20 possible choices, 20 different amino acids that could potentially appear at that link. So for a chain of 3 amino acids, there are 20 times 20 times 20 = 20^3 = 8,000 possible combinations; for a chain of 300 amino acids there are 20^{300} (about 10^{390}) possible combinations. So if a chain of 300 amino acids were formed by accident, the odds that it would be a specific working protein would be 1 in 20^{300}. 20^{300} is a fantastically large number, and 1 in 20^{300} is a fantastically small number. For comparison, the entire visible universe contains about 10^{80} protons, neutrons, and electrons.

In general, there is some flexibility, some substitutions of amino acids that will result in a protein with the same function—that works the same—in a living cell. In the late 1980s, MIT biochemist Robert Sauer estimated the probability that a sequence of 92 amino acids would result in a functional protein to be 1 in 10^{63}.[22] Douglas Axe later studied one protein of

150 amino acids and estimated the odds of obtaining a functionally equivalent protein in the space of all possible 150-unit sequences of amino acids as 1 in 10^{77}.[23] In other words, if you take a protein made up of 150 linked amino acids and you randomly change or "mutate" its amino acids, the odds that the resulting protein will be able to perform the same function are about 1 in 10^{77}. (Axe also estimated that the probability of getting any type of functional protein out of linking 150 amino acids as one in 10^{74}.)

These probabilities are incredibly small. Consider the odds of linking 150 amino acids and getting a specific working protein, which Axe estimates is about 1 in 10^{77}. If the entire Earth were covered 2 miles deep with the exact perfect 20 amino acids for life, and those amino acids were somehow forming themselves into perfect 150-unit chains every millionth of a second since Earth was formed (around 5 billion years ago), the probability that our specific working protein would ever arise, even for one millionth of a second, is still less than one in a billion.* What is at work here is the power of exponents. (For those not familiar with exponents, appendix A gives an introduction.)

There is a legend in India that Lord Krishna once appeared as a sage in the court of a king. The king loved to play chess, and asked the sage to name his own prize in the unlikely event he could beat the king. The sage (Krishna) asked for 1 grain of rice on the first square of the chessboard, then 2 grains on the second square, 4 grains on the third square, and so on, with each square having double the rice on the preceding square, up to the sixty-fourth square. The king thought this small prize was not worthy, but he agreed. The king lost, and proceeded to pay. By the twentieth square, there were over 1 million grains of rice, and by the fortieth square, there were over 1 trillion grains of rice. The king realized he could not pay. To pay, he would have needed about 1.8×10^{19} grains of rice, or about 18 billion billion grains of rice.†

So doubling—multiplying by two—sixty-four times takes us from a single grain of rice to 18 billion billion grains of rice. To get the odds that a specific working protein of 150 amino acids was "accidentally" created,

* This is derived in part from estimating the volume of a typical amino acid at 100 cubic Angstroms, or 10^{-28} cubic meters, and the surface area of Earth at about 5 times 10^{14} square meters.

† See http://www.singularitysymposium.com/exponential-growth.html. According to the legend, Krishna told the king he could pay his debt over time, by feeding visitors.

even assuming we have a chain of 150 linked amino acids, we have to multiply by ten times ten times ten and so on for a total of 77 tens.

Exponents are overwhelming. With exponents, repeated multiplication, numbers get very big very fast. The number of possible proteins— of possible chains of 20 different amino acids linked end-to-end—is staggering, even for relatively short chains.

An analogy I found in the literature, and that has a somewhat comical appeal to me, is that of a blind monkey typing. Suppose that, instead of 26 letters plus spaces and punctuation in our English alphabet, there are only 20 possible choices for the monkey for each keystroke. We'll use 20 because that's the number of different kinds of amino acids our cells link together to make proteins (for now, we'll ignore substitutions of amino acids; the math is so overpowering it works either way, and some proteins allow little or no substitution). We can compare the odds of a monkey accidentally typing a specified sequence of letters with the odds of the accidental formation of a specified sequence of amino acids.

Here's a line from Shakespeare: "What fools these mortals be."* It has a total of 27 letters/spaces. How likely is it that a monkey will accidentally type this exact sequence? Again, we assume the monkey has a choice of exactly 20 letters/spaces for each keystroke and is equally likely to type, completely at random, any particular letter/space in any position. Of course, our blind monkey could get lucky and type this exact sequence the very first time. But what are the odds?

Because there are 27 letters/spaces, the number of possible sequences of this length is 20^{27}, or about 10^{35}. The number of seconds in the history of the universe, since the big bang, is about 4×10^{17}. So even if we have a billion (10^9) quick-typing monkeys, who type the entire 27 characters every single second since the beginning of the universe (obviously this is fanciful, the monkeys couldn't have existed at the beginning of the universe, but we are just trying to get a sense of the magnitude of the problem), we "only" get a total of 4×10^{26} sequences, and the chance that any of these monkeys ever typed "what fools these mortals be" is about 1 in 250 million (because 10^{35} is about 250 million times bigger than 4×10^{26}).

So the odds of accidentally forming a precise sequence of 27 amino acids are laughingly small. Let's now expand our imaginations to assume

---

* The quote is attributed to Roman philosopher Seneca the Younger.

that every star in the universe has a planet with 1 billion monkeys typing. If we use a large estimate for the number of stars in the universe, we can expect that one of these monkeys will type "what fools these mortals be" within about 10 seconds.*

But that's just the beginning of complexity. Now let's estimate the odds of accidentally creating a specified string of 75 amino acids. There are 75 letter/spaces in "What a piece of work is a man how noble in reason how infinite in faculties." There are a total of 20^{75} or about 4×10^{97} amino acid sequences of this length. Suppose, instead of a mere billion monkeys, each star in the universe has a planet with a billion billion billion billion monkeys, or 10^{36} monkeys, and each monkey can type even this 75 letters/spaces sequence in just 1 second, to keep it simple.† OK, maybe too much math, definitely too many monkeys, but the likelihood of any of these blind monkeys ever typing our specified sequence in a time equal to the history of the universe is less than 1 in 10^{19}, or less than one part in ten billion billion. So the odds of forming a single, relatively short specified protein with 75 amino acids by accident are vanishingly small, regardless of how many stars with Earthlike planets may exist. We are beyond laughingly small to pretty darn ridiculous.

I apologize to those who find the math tedious, but here's the basic point, in plain English. Don't tell me amino acids can be created by accident. Don't tell me about "billions and billions" of years for life to arise. Don't tell me about "countless" stars and planets in the universe. It all doesn't matter. Using simple concepts of number—exponents—one can expose as false claims that life arose by accident. You cannot seriously expect to get a specified protein of 75 linked amino acids in the history of the universe, except as a product of already existing life, even if you assume that everything in the universe is made up of amino acids and even if you assume that amino acids will freely combine into 75-unit chains.‡ Period. And there actually is no dispute about this fact.

---

* Some current estimates are about 10^{23} stars in the universe, but let's say 10^{25} to be generous. So this calculation assumes 10^{34} imaginary monkeys typing!

† Now, over the life of the universe, all of these monkeys can be expected to type about 4×10^{78} sequences. For this calculation we multiply 4×10^{17} (the number of seconds since the big bang) times 10^{25} (number of stars) times 10^{36} (number of monkeys on each star), to get 4×10^{78} sequences.

‡ As noted, amino acids do not naturally link together. It generally takes energy and special machines in life to "snap" amino acids together.

Of course it takes more than one functional protein to create life. If you use blue-green algae as a model for the first life-form, it takes perhaps 2,000 functional and exquisitely coordinated proteins. Do all of the amino acids in all of the proteins have to be left-handed (see chapter 11)? Perhaps, but let's ignore that for now. You also need molecules that will store the exact code needed to build those proteins. Some of those proteins will be the machine parts that will read the code and assemble other proteins. And the code itself has to be exactly in the right order to build proteins with amino acids in the right order.

Nucleotides—the groups of atoms that hold the code for life in DNA—are impossibly unlikely to form by chance. One scientist estimates the odds of generating a batch of nucleotides by accident as one in 10^{109}.[24] And that's a batch of nucleotides in random order. To put them in a specified order, in the exact right order to build the 2,000 or so proteins for primitive life, is mind-boggling unlikely.

Here's a cute quote from former Atheist Fred Hoyle:

> Imagine 10^{50} blind persons each with a scrambled Rubik's cube, and try to conceive of the chance of them all simultaneously arriving at the solved form. You then have the chance of arriving by random shuffling, of just one of the many biopolymers on which life depends. The notion that not only the biopolymers but the operating program of a living cell could be arrived at by chance in a primordial organic soup here on the Earth is evidently nonsense of a higher order.[25]

The origin of life really is a chicken and egg problem. To get life, you need BOTH the exact code and the proteins—machine parts—to read the code and build new proteins, and to do all the other wonders of life. Some of the code gives instructions on how to build the proteins that read the code. How could it all have simultaneously arisen?

I focus in this book on the DNA/protein enigma, but life's chicken and egg problem is more serious. Here's biologist Ann Gauger:

> Everything in organisms is interconnected causally. Everywhere in biological systems, chicken and egg problems abound. For example, amino acid biosynthesis pathways are composed of enzymes that require the amino acids they make, ATP biosynthesis pathways must have ATP to make ATP, DNA is needed to make proteins, but proteins are needed to make DNA, and the list goes on. Indeed, the scope of the problem is difficult even to grasp.[26]

Let me give you another example. Suppose you were in a magic building. You pick a quality, and every time you go up one level, it increases by a factor of ten, and every time you go down one level, it decreases by a factor of ten. Suppose you pick length to start, and you start with one meter. If you go down just four levels, you have 10^{-4} meters, which is the average width of a human hair. When you get down eleven levels, you have less than the radius of a hydrogen atom; down eighteen levels, and you have the size of a single electron; down twenty-four levels, you reach the cross section of a neutrino; and below that level, you quickly get into what physicists call the "quantum foam"—a scale at which ordinary concepts of matter and space do not apply.

If you again start with one meter and go up, by the eighth level (7 levels up), you've got the length of China's Great Wall; by the tenth level, the diameter of the Sun; by the seventeenth level, the outer boundary of the Oort Cloud of frozen comets around the Sun; and by the twenty-eighth level, the entire distance across the visible universe. So about 24 + 28 = 52 levels are enough to span all known lengths in the universe, to go from the distance across the universe to a distance so small that ordinary concepts of matter do not apply.

Now you pick probability, and you start with the probability 1, the expectation that something will happen an average of once in the time span you are observing. If you go down 9 levels, you get a probability lower than your expectation of winning any Powerball or Mega lottery. If you go down 27 levels, you have the probability of picking at random a specified or marked toy marble out of a pile of marbles as big as the entire Earth. If you go down 150 levels, you get to Bill Dembski's "universal probability limit" (see chapter 6), below which specified events or specified patterns cannot reasonably be attributed to chance. To get the probability that life ever formed by accident anywhere in the universe, you have to go down at least tens of thousands of levels, and perhaps millions or even billions of levels.

Yale physicist Harold Morowitz calculated the likelihood of life arising by chance as one in 10 to the one hundred billionth power (one in $10^{100,000,000,000}$).[27] Going down one hundred billion levels in a building would be like coming out on the other side of Earth after 8,000 miles of descent, and then continuing twice as far as Earth is from the Sun. This in a building where 52 levels span all known lengths in the universe.

Life forming accidentally is like a tornado ripping through a massive junkyard and leaving behind a 747 jet, with all systems functional and ready to take off. Except it's worse. The tornado would also have to leave behind a complete set of blueprints for building the jet and an operating manual.

So was life created by accident? Relatively straightforward mathematics says not likely. Unbelievably, unimaginably, unlikely. It doesn't matter whether you "just" have to go down 10,000 levels in that magic building or one hundred billion, it's out of sight either way. That's based simply on science and number, not on any appeal to religious belief. Many Atheists agree that life could not have formed by accident alone. It's not a religious statement in any way, although there doesn't seem to be much of a rush to update those high school textbooks.

Such relatively simple calculations shocked me. I wanted to share, and that desire ultimately became this book.

An Atheist might respond to all of the facts of this chapter by saying that, because we're here, it doesn't really matter how unlikely it was for life to get started. In a way that's true. You can shut out the wonder, and you can close your mind to the possibility of design and the existence of God. But to me, the discovery of the complexity of life, and the inability of modern science to conceive of even a plausible pathway for its existence, is a wonder.

We now have three pillars of modern science supporting Belief—the creation of the universe, the fine-tuning of the universe, and the origin of life. We have counted to three. Let's look next at the technology of life. That is a wonder by any standard.

CHAPTER 11

The Technology of Life

How does life work?

DNA is biological computer code, only far, far more advanced than anything we have ever built.

BILL GATES

LIFE IS A TECHNOLOGICAL miracle. Life has a digital operating system, futuristic information input and storage technology, information retrieval machines with built-in proofreading, splicing technology, molecular delivery trucks, factories to build needed machines and parts, technology to compress information, shredding technology, replication technology, repair technology, and so much more. We'll look at these and then return briefly to probability, to calculate the odds that just one small piece of functional nanotechnology was created by blind chance. To me, the fantastic, futuristic technology now known to exist in all life, and the overwhelming improbability that even a small piece of this was created by blind chance, are the fourth wonder of modern science, the fourth of seven in our count to God.

In Charles Darwin's day, 150-plus years ago, the technology of life was unknown. Early microscopes revealed the existence of cells, but they showed little else. The stuff inside cells was called "protoplasm," and it was thought to be a jellylike goo that somehow did all of life's work. Nobody knew how or why. Cells were "homogeneous globules of plasm."[*] I think if Charles Darwin were alive today and could study the technology now known to exist in all life, he would agree that all of this technology could not have arisen by pure chance.[†]

Your body has about 30 trillion cells, that's 30,000,000,000,000 cells.[‡] Thirty trillion is a ridiculous number. You've got about 100,000 cells for every person in the United States. If our Milky Way Galaxy has 400 billion stars, you've got about 75 cells for every star in our galaxy.

Each of your 30 trillion cells is like a complex, three-dimensional city, with libraries, factories, and highways. Here's biologist Michael Denton:

> To grasp the reality of life as it has been revealed by molecular biology, we must magnify a cell a thousand million times until it is twenty kilometers in diameter and resembles a giant airship large enough to cover a great city like London or New York. What we would then see would be an object of unparalleled complexity and adaptive design. On the surface of the cell, we would see millions of openings, like the portholes of a vast space ship, opening and closing to allow a continual stream of materials to flow in and out. If we were to enter one of these openings, we would find ourselves in a world of supreme technology and bewildering complexity. We would see endless highly organized corridors and conduits branching in every direction away from the perimeter of the cell, some leading to the central memory bank in the nucleus and others to assembly plants and processing units. The nucleus itself would be a vast spherical chamber more than a kilometer in diameter, resembling a geodesic dome inside of which we would see, all

---

[*] This quote is attributed to German biologist Ernst Haeckel (1834–1919), who advanced Darwin's theories. See http://www.discovery.org/a/1764.

[†] Darwin's great-great-great granddaughter Laura Keynes has stated that her faith was restored through intellectual pursuit. See James Kelly, "If Only Charles Darwin Could See His Descendant Now." National Catholic Register, August 14, 2013, http://www.ncregister.com/daily-news/if-only-charles-darwin-could-see-his-descendant-now?utm_source=feedburner&utm_medium=feed&utm_campaign=Feed%3A+NCRegisterDailyBlog+National+Catholic+Register#When:2013-08-14%2014:32:01 (accessed September 2, 2013).
An excellent book—I, Charles Darwin by Nickell John Romjue (Tucson, AZ: Whatmark, 2011)—imagines how Charles Darwin might react if he were to return to Earth in 2009, the 150th anniversary of the printing of his book On the Origin of Species.

[‡] This book uses 30 trillion as the number of cells in a human body, but I was surprised to learn that estimates vary widely, from 10 trillion up to 100 trillion. It doesn't matter much for purposes of this book, and 30 trillion seems to be the number most commonly given. I am ignoring here nonhuman, bacterial cells; we each have around 100 trillion of those. Bacterial cells are smaller.

neatly stacked together in ordered arrays, the miles of coiled chains of the DNA molecules. A huge range of products and raw materials would shuttle along all the manifold conduits in a highly ordered fashion to and from all the various assembly plants in the outer regions of the cell.[1]

Your 30 trillion cells are connected by and controlled through a complex communication system. This system tells them when to divide and, in some cases, when to self-destruct and die. It tells them when and how to work together, like the muscles in your body. Muscle cells contain a protein called titin, which works like a spring. Titin is the largest human protein, 34,350 amino acids linked together. Consider the programming needed to coordinate billions of cells in different muscle groups so that these machines move your body at your will. Consider the technology needed to coordinate your muscles with your eyesight so that you can, with relative ease, catch a ball, or lift your foot over a rock.

I see in the technology of life an absolute wonder. I think if we as a society could somehow see it untarnished, free of the Scientism stranglehold, we would rejoice at the miracle and bow our heads to the skill of a master designer. But few see it at all, and most who do have been "educated" to ignore the obvious. This cannot stand. A time will come, be it ten years or a hundred years from now, when the truth will escape its academic shackles, when the paradigm breaks and the wonder sparkles. Life is not some protoplasmic goo. Life is a technological miracle. Each year reveals greater complexity, technology to amaze. Perhaps more than in any other area of science, in molecular biology the arrow of scientific progress points directly to a master designer, to God.

This was the hardest chapter to write, because the wonder here is overwhelming. I'm going to focus on life's technology relating to biological information. Between graduate school in theoretical mathematics and law school I worked for almost two years as a computer programmer. I developed a heightened appreciation, from hundreds of mistakes, of how the code you put into the computer has to be exactly right.

This chapter will vastly oversimplify the technology of life. It could be thousands of pages long (but don't despair, it's not). We won't discuss evidence that many mutations are "directed mutations," where programming in the cell mutates the organism as needed and the organism rewrites its own DNA code.[2] We won't discuss evidence that organisms store substantial information outside their DNA code.[3]

To me, the discoveries of the technology of life are the fourth wonder of modern science. Put on your seat belt, we have only begun. I'm going to hit you with a lot of facts, because here God is revealed in the details. If one of the sections below seems dense, feel free to skip it and go to the next. There will be no quiz. But if you have the time, and the interest, I promise you wonder aplenty.

The Central Dogma—Life's Digital Operating System

When I was a computer programmer, I used a language known as FORTRAN. I "punched" computer cards, each with the precise code for one "line" of instructions. I would submit my batch of cards for white-coated technicians (the only people allowed in the room with the huge IBM mainframe) to enter into the computer. A few hours later, more or less, I would get the results, which were usually gibberish, because even though I thought of myself as a good programmer, it was hard to write perfect instructions. It was mostly "garbage in, garbage out." I didn't know how the computer actually processed (or choked on) my FORTRAN instructions, but I knew the computer had a sophisticated operating system.

An operating system is a set of instructions (software) that control machine parts (hardware). To get the machine parts (computer, printer, monitor, etc.) to do what you want, you have to enter information in the form specified by the operating system, information in the form the operating system is designed to recognize. Human beings have designed a variety of operating systems, and they generally change and get more sophisticated over time. Operating systems for copying and playing back sound—from records to 8-tracks to CDs to MP3 files—are a good example. CDs and MP3 files are digital systems, because the information comes in discrete bits, not in a continuous flow (like older analog TVs or records). FORTRAN is a digital language; I typed discrete letters, numbers, and symbols on cards to be read. A digital operating system has a code. It may be binary—0s and 1s—or it may be more complex.

All life has the same advanced digital operating system.* All life stores its code in long molecules of DNA. All life has machines that transfer— transcribe—DNA information to a slightly simpler molecule called "RNA." All life has machines that process the information in RNA to build proteins and other essential molecules of life. This information transfer system—DNA to RNA to proteins—is called the "central dogma" of molecular biology.† It explains how life stores and processes information. It is the operating system of life.

How can it be that all life uses the same operating system, runs off the same software platform? The wonder of this fact is apparent from the following excerpt on page 2 (out of 1,600) of the fifth edition of *Molecular Biology of the Cell,* a standard college textbook:

> Living cells, like computers, deal in information, and it is estimated that they have been evolving and diversifying for over 3.5 billion years. It is scarcely to be expected that they should all store their information in the same form, or that the archives of one type of cell should be readable by the information-handling machinery of another. And yet it is so. All living cells on Earth, without any known exception, store their hereditary information in the form of double-stranded molecules of DNA—long unbranched paired polymer chains, formed always of the same four types of monomers. These monomers have nicknames formed from a four-letter alphabet—A, T, C, G—and they are strung together in a long linear sequence that encodes the genetic information, just as the sequence of 1s and 0s encodes the information in a computer file. We can take a piece of DNA from a human cell and insert it into a bacterium, or a piece of bacterial DNA and insert it into a human cell, and the information will be successfully read, interpreted, and copied.[4]

Did you catch that last point? It shocked me, and this is just page 2. (If you want to make sure you are recognized as a nerd, just walk around holding this seven-pound book.) "Simple" one-cell organisms all have molecular machines that can read and process DNA from a human cell. The vast majority of the 30 trillion cells in our bodies have similar machines that can read

---

* In the 1940s John Von Neumann, who is sometimes called the father of modern computing, proved that an entity that can reproduce itself must store all the information for that entity, including how to reproduce, in some sort of code.

† The phrase "central dogma" was first stated by Francis Crick when he proposed this information transfer theory in 1958. Crick discussed the phrase in a paper he published in August 1970, "Central Dogma of Molecular Biology," *Nature* 227, no. 5258: 561–63.

and process DNA from bacteria. The code of DNA has the same form and is read the same way for all life. Sure the order of the letters and the length of the DNA code are different from organism to organism, much like the order of the letters and the length of the text are different from one book to another. But just as the English language has a standard alphabet of twenty-six letters, the code of DNA has a standard alphabet of four letters, and all life reads that alphabet the same way (three letters at a time) to build proteins. The information of life comes in digital form (discrete bits of information), and that encoded, digital information is processed the same way for all life.

There is no scientific evidence of a different operating system for life. It is quite possible the operating system of life sprang into existence with all of its astonishing complexity 3.5 billion years ago, and has not changed since that date.* To me, that is profound. We are bombarded with what I view as unscientific propaganda about the power of evolution—how evolution created everything, shapes everything, and constantly refines everything. The college textbook containing the quotation above, which in a crude sense is sort of the "bible" of molecular biology, repeatedly claims that evolution did this or refined that. Yet you can sense the astonishment of even these worshippers of neo-Darwinian theory that the operating system of life has not evolved, that it appears somehow, inexplicably, to have arisen fully formed, fully developed, and fully sophisticated. To me, this basic fact—on page 2 of the primary textbook on molecular biology—is blatantly inconsistent with the prevailing scientific paradigm that evolution created everything through natural selection.

It's undisputed. All life has the same operating system. There is no evidence there ever was a different operating system—if there had been, we would expect to find that it "evolved" into different operating systems today.

Think of it in the language of modern technology. Compare human beings, with 30 trillion exquisitely interconnected cells, to the most primitive bacteria. We have astonishing added hardware: legs, arms, hearts, eyes, ears, and on and on. We have amazing "apps": all of our senses, muscle coordination, subconscious regulatory processing, and the killer

---

* Estimates for the origin of life on Earth vary. Some chemical evidence suggests 3.8 billion years ago or even earlier; some scientists suggest 3.4 billion years ago or a more recent date. For our purposes, it doesn't matter much, and this book will use 3.5 billion years ago as the estimated date for the origin of life.

"app" of all existence: human consciousness and reason. Yet we have the same operating system as the bacteria. We and that bacteria and all life run off Life 1, designed 3.5 billion years ago.

This undisputed fact is a problem for Darwinists, and they seek to blunt it by proposing an RNA world before the development of DNA. RNA has a simpler "backbone"—one sugar unit instead of two—and they propose a single strand of code instead of the double helix. But the RNA world proposal has major flaws, as noted by Stephen Meyer in *Signature in the Cell*.[5] RNA building blocks break down in water. RNA building blocks are hard to synthesize and easy to destroy. RNA molecules can perform only a few of the thousands of functions performed by proteins. There is no known way RNA could have evolved simultaneously into DNA and the hundreds of proteins necessary to run life's operating system.

The RNA world proposal derives from philosophic discomfort, not science. RNA world proponents are like the steady state proponents of the mid-1900s—a group of scientists (Fred Hoyle was one) who invented a theory of matter being created out of nothing to blunt the philosophical implications of Hubble's law and the creation of the universe in a single big bang. There was no scientific evidence that matter could be created out of nothing; the steady state theory blatantly violated the first law of thermodynamics, the law of conservation of mass and energy. Similarly, today's RNA world proponents disregard scientific flaws to propose—with no evidence—a simpler coding structure in the origin of life. They do this solely because of their philosophic discomfort with the scientific evidence that the incredibly elegant DNA code, and all of the sophisticated machinery to process it, were fully formed when life began.

The RNA world proposal does not explain where the information to build any operating system came from. It cannot evade probability. How, by accident, could there have arisen, simultaneously in the same confined space, (1) thousands of complex proteins and other molecules perfectly assembled to build chemical machines and perfectly designed to work with each other to read, process, and copy coded information; and (2) the exact correct code to build every one of those thousands of machine parts? Keep in mind it is highly improbable a single one of these proteins could ever form by accident in the entire history of the universe. Darwinist belief in the accidental creation of any operating system is a fanatical belief. It is contrary to all the facts and evidence of science. The evidence of modern science favors design, and it is not a close call.

Enough rambling about a 3.5-billion-year-old operating system. Let's look at the futuristic information input and storage technology it comes with.

DNA—Futuristic Information Technology

As a programmer, I typed information on cards to be read by the computer. Life reads information on molecules of deoxyribonucleic acid, or DNA. Let's see how modern science solved the riddle of DNA.

Experiments in the early 1940s suggested DNA contained the information of life, the instructions for life. When DNA of smooth, lethal bacteria was removed, purified, and added to similar rough, nonlethal bacteria, the rough bacteria were transformed into smooth, lethal bacteria. Scientists concluded that somehow the information to build smooth bacteria was contained in DNA. But exactly what was DNA, and how did it store and pass on information?

In the early 1950s, scientists looked for answers. Chemical evidence suggested that DNA had one or more backbones of linked sugar and phosphate molecules, and four groups of atoms somehow attached to those backbones. DNA was known to be a stable molecule. There was growing suspicion, and growing X-ray evidence, that DNA was a spiral molecule and that the basic structure essentially repeated itself, by completing one loop of the spiral every 34 Angstroms (an Angstrom is 10^{-10} meters). There was also puzzling evidence that the four groups of atoms attached to DNA (called "nucleotides") always paired up in amounts; the amount of adenine (A) always equaled the amount of thymine (T), and the amount of cytosine (C) always equaled the amount of guanine (G).

So the race to solve the riddle of DNA was on. A leading contender was American chemist Linus Pauling. In 1952 Pauling thought he had won. He proposed a center of three sugar phosphate backbones with the nucleotides on the outside.

Pauling described his DNA model in a letter to his son Peter, who was at Kings College in Cambridge, England. Peter shared a lab with a somewhat eccentric pair of scientists who were also working to solve the DNA riddle. James Watson was a young, basically unknown American scientist who was unsure what area to specialize in. Francis Crick was a fast-talking, equally unknown Englishman still working on his PhD.

Watson and Crick had the particular good fortune of access to the work and theories of Rosalind Franklin. In 1952, she took a famous X-ray diffraction image of DNA, known as "Photo 51." James Watson obtained the photo without Franklin's knowledge or permission. Franklin discussed her theory—in lectures attended by Watson and in reports accessible to Watson and Crick—that based on X-ray evidence such as Photo 51, DNA appeared to be a double helix with antiparallel strands and with the phosphate backbone on the outside.

Photo 51 has been described as the "critical data" in the discovery of the DNA double helix. Calculations from Photo 51 helped determine DNA's size and structure. In recent years, Rosalind Franklin's vital role has been rediscovered. Her story, "The Secret of Photo 51," is told in an episode of *NOVA*. Because she died from cancer at age thirty-seven, four years before the Nobel Prize for the discovery of DNA was awarded and because you cannot be nominated for a Nobel Prize after you die, Rosalind Franklin did not share in the prize, and so at the time her enormous contribution was barely noted.

When Watson and Crick read Peter's letter from his father, they knew, from their exposure to the work of Rosalind Franklin, that Pauling's model was not correct. They also knew Pauling would soon discover his mistake. They pressed the machine shop to make models of molecules. They put the models together with a double sugar-phosphate backbone on the outside. They first thought the nucleotides on the inside should pair up like-to-like—that is, each backbone should have the same one of the four nucleotides at each position. But the size and shape of the four like-like nucleotides pairs varied significantly, and they could not find a stable model in which the sugar-phosphate backbone could wrap around them.

Playing with the models, James Watson noticed that adenine (A) would form hydrogen bonds with thymine (T) and cytosine (C) would form hydrogen bonds with guanine (G), and that, even more significant, the resulting bonded pairs were of very similar shape and size. Within days, Watson and Crick put together a DNA model with two outside spiral backbones, going in opposite directions, holding A-T and C-G pairs. Their model matched the X-ray evidence and the theories of Rosalind Franklin. In 1953, they published their theory of DNA design. In

1963, they were awarded the Nobel Prize. James Watson's account of their work, *The Double Helix,* became a bestseller.

DNA is the superstar molecule of life. We now know DNA holds the instructions—the "code"—for all life. Every living thing—every bacteria, fungus, plant, and animal—has in the majority of its cells an enormously complex DNA molecule or molecules, that specify, by a precise sequence of the four nucleotides, the code to build that organism. The illustration below shows the structure of DNA and the related molecule RNA.

According to Michael Denton, "DNA and RNA may be uniquely fit for their respective biological roles."[6] DNA is small, because the code is written with groups of atoms. When an item gets smaller in each dimension, the space needed to hold it goes down dramatically, by the cube of the shrinkage. The letters of DNA are less than one-millionth the size of the letters in this book.

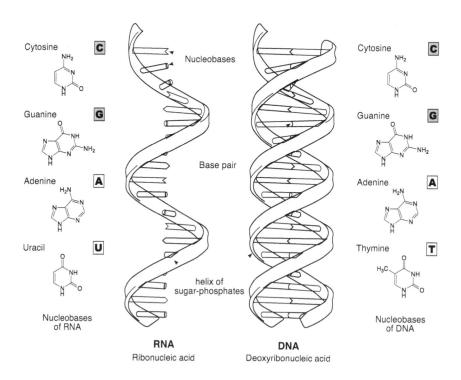

DNA is a fantastic technology for storing information:

> Scientists have been eyeing up DNA as a potential storage medium for a long time, for three very good reasons: It's incredibly dense (you can store one bit per base, and a base is only a few atoms large); it's volumetric (beaker) rather than planar (hard disk); and it's incredibly stable—where other bleeding-edge storage mediums need to be kept in sub-zero vacuums, DNA can survive for hundreds of thousands of years in a box in your garage.[7]

The above article, from August 2012, reported that scientists were able to store information in DNA, so compact that a single gram could contain 700 terabytes of data, or the equivalent of "14,000 50-gigabyte Blu-ray discs . . . in a droplet of DNA that would fit on the tip of your pinky." Michael Denton calculated in 1996 that a mere spoonful of DNA could store the complete instructions for every species that has ever existed, with enough room left over to store every book ever written.[8] Scientists are now working to store information in synthetic DNA.[9] How ironic that DNA, the dream information storage technology of our future, is 3.5 billion years old.

Your DNA is the code used to build you. It takes a lot of DNA to build a human being, about 3.2 billion subunits—"letters"—of DNA. Almost every one of the 30 trillion cells in your body has forty-six very, very long molecules of DNA, called "chromosomes."* On average, there are 70 million DNA code subunits in each of your forty-six chromosomes, linked end-to-end like an impossibly long series of plastic connecting blocks. Your 3.2 billion letters of DNA would fill over a thousand thick volumes with fine print. When I was a computer programmer and wanted to give the computer instructions to perform a task, I typed dozens or hundreds of computer cards. Your DNA holds the instructions to build and operate you in 3.2 billion letters of information.

The DNA strands in your 30 trillion cells are about 2 billionths of a meter in diameter. If the DNA in just one of your cells were magnified 5 million times, so that it was just 1 centimeter in diameter, like light rope, and laid end to end, it would measure 6,000 miles, or the distance from

---

* Red blood cells have no nucleus and thus no DNA, except for the DNA in mitochondria. Reproductive cells (gametes) have only twenty-three chromosomes. Cells also contain mitochondria, specialized compartments that produce energy the cell can access and that contain their own DNA. Mitochondrial DNA is passed down only from maternal ancestors.

Los Angeles, California, across the entire Pacific Ocean to Seoul, Korea. At that magnification, a typical cell would have the size of a sphere about 800 feet in diameter. And inside that sphere would be 6,000 miles of light rope, very carefully coiled and stored.

A subunit of DNA is a piece of backbone plus the chemical group for a single letter of DNA. Strong "phosphodiester" bonds (centered by a phosphate atom) hold the backbone parts of the subunits together, and the strong outer backbone shields the code of nucleotides. Each subunit has a direction to it, again sort of like a plastic connecting block with a knob at one end. Now imagine 70 million of these blocks—DNA subunits—snapped together to form just one strand of one of your forty-six chromosomes. Because the strands point in opposite directions, the biological machines that work on DNA can tell the strands apart. Because the chemical bonds between the backbones of the subunits are much stronger than the chemical bonds between the interior nucleotides, the letters of DNA, these biological machines can "upzip" your DNA and read it or copy it.

Just as, in a book, any letter can appear in any space, so too in DNA any chemical letter can appear at any place. DNA is the code for life, and the code is not generated by any physical or chemical property. In language, there are rules of spelling and grammar that tell us what combinations make sense, and DNA has roughly analogous rules, but with both a book and DNA, the creative process starts fresh, with a blank slate.

This proves life is not an inevitable or automatic consequence of the laws of nature. Just as there is no natural law that compels the existence of a work of Shakespeare, so too there is no natural law that compels the "writing" of working DNA—a working code for life.

I started my journey with a vague sense that DNA was some sort of code. I learned that, in the words of Bill Gates, "DNA is biological computer code, only far, far more advanced than anything we have ever built."

Information Retrieval Technology

Information retrieval technology is in all life. The evidence suggests it existed when life was first created.

As we have seen, DNA stores information that cells use to build the parts they need. These parts are mostly proteins, and the stretch of DNA that contains the code to build a particular protein is called a "gene." The central dogma tells us the path from DNA to protein has an intermediate step. The code in the DNA is read and copied—transcribed—into an intermediate information molecule called "messenger RNA." So the three steps of the central dogma—life's operating system—are (1) DNA to (2) messenger RNA to (3) protein. Let's look at how cells go from step one to step two.

To read DNA, the double helix has to be opened up to exactly the right section and unzipped so that a single strand can be read. Think of going to the library and making a copy of a page in a book. If you want to take information out of a library but can't take the book out, you go in, find the right book, open it to the right page, copy the information you need, put the book back together, and then take the information outside. The process of reading—"transcribing," or retrieving information from—DNA works roughly the same way.

Let's talk machinery. All life has in each cell an incredible machine called "RNA polymerase." RNA polymerase moves along DNA, unwinds it, unzips the double helix, copies the four-letter code, and puts the double helix back together. It has over one hundred parts, which are proteins.

RNA polymerase is life's information retrieval machine. It captures the information in a single strand of DNA and loads it into a created molecule of messenger RNA. Messenger RNA is a single strand of information; it is basically a copy of the chemical code—the four chemical letters A, C, T, and G—of DNA. (One of the letters is chemically altered in messenger RNA, but it doesn't affect the analysis, and we will ignore that detail going forward.) RNA stands for "ribonucleic acid," and it is very similar to a single strand of DNA, although its backbone is less strong. The biological term for this information retrieval/copying process is "transcription."

RNA polymerase builds messenger RNA precisely. On average, it makes about one error in every 10,000 DNA letters. If you were transcribing a book with this accuracy, there would be on average one letter wrong about every five pages. RNA polymerase has a built-in proofreading mechanism—in some cases it backs up, cuts out a bad letter or section, and starts transcribing again.

Because DNA is a helix that makes a complete turn about every ten letters, there is a coiling problem. The DNA of all life gets twisted as it is unwound. Another special biological molecular machine solves this coiling problem. It takes two strands of DNA, cuts a complete break in one of them, passes the uncut DNA strand through the cut strand, and then reseals the cut DNA strand. It does this exactly as needed. All of this sounds incredibly fantastic; modern science proves it to be true.

The process is more complex in eukaryotic cells, cells with a nucleus. The step from bacteria to cells with a nucleus may have been a greater step in complexity than the first step to create life. An average mammalian cell is two thousand times larger than the bacteria *E. coli*.[10] Bacterial DNA is contained in a geometrically simple loop (although not a short loop, the DNA of *E. coli* has 4.6 million letters). In cells with a nucleus, DNA is coiled on spool-like combinations of proteins called "nucleosomes" (not to be confused with "nucleotides," which are the groups of atoms that contain the letters of DNA). To retrieve information from a eukaryotic cell, the RNA polymerase has to move through the complex "chromatin" structure (discussed in more detail below), unwind the DNA off the nucleosome spools, unzip the double helix, manufacture a molecule of RNA with the same chemical code, and reconstruct the DNA—zip it back up, put it back on the nucleosome spools, and put the chromatin fiber back together. Eukaryotic cells have three different types of RNA polymerase machines.

Splicing Technology

Human beings and other advanced forms of life come with splicing technology. After RNA polymerase retrieves the information from a section of DNA to create a long molecule of messenger RNA, other machines cut out and "splice" the good sections together in just the right way to construct a finished product that is the code for whatever protein that cell has decided to build. Many genes can be spliced in different ways to build different proteins. This technology reduces the DNA code needed. About 75 percent of human genes have this ability, called "alternate splicing," to encode more than one protein.*

The splicing machinery requires five RNA molecules and over two hundred proteins[11] Some of the sections that have to be sliced out can be as long as 100,000 RNA letters. So the splicing machine—called a

---

* This ability to encode the information for more than one protein is valuable. In May 2012 scientists announced that the tomato has 31,760 genes, about 7,000 genes more than humans. However, tomato genes generally encode the information for a single protein. See *New York Times*, May 30, 2012, Science section.

"spliceosome"[12] (I really like that name)—has to be extremely accurate. It contains its own proofreading machinery. Exactly how it all works is not fully understood.[13] It is also unclear how the spliceosome knows when it has reached a "good" section of DNA where the copied letters must be retained in the RNA.

Titin, that springlike protein in your muscles made from 34,350 amino acids linked together, has a molecular weight equal to 3.8 million hydrogen atoms. To get the code to build titin, your spliceosomes cut out and carefully splice together 363 sections of messenger RNA. This is science fact. The theory that all of this arose by chance is science fiction.

Molecular Delivery Trucks

Think of all the steps and all the information it takes to get a package to your front door. Cells with a nucleus have technology to escort finished RNA through the membrane of the nucleus. Special proteins—technically called "nuclear pore complexes"—attach to finished RNA molecules, transport them through the membrane of the nucleus, deliver them to factories called ribosomes, release the RNA, and then reenter the nucleus to transport more RNA.* They are a molecular shuttle service, biological trucks.

Questions?

At this point, you may have questions. I sure did. How does a cell know when it needs to build a particular machine part—a particular protein? How does a cell know where to find the instructions (gene) for that particular machine part/protein in your 3.2 billion letters of human DNA? After the gene is copied onto messenger RNA, how does the cell know exactly which sections of that messenger RNA are "good," and how does it cut out only the good sections and know how

---

* Only a small fraction of the RNA that is created in the nucleus is "finished" RNA that can be used. The rest—such as spliced-out sections of RNA and damaged or faulty RNA—is not used to code for proteins, although some is used for other purposes. Special proteins prevent useless RNA molecules from being transported outside the nucleus. These proteins apparently identify useless RNA for molecular machines called "exosomes" that roam the nucleus and recycle RNA. Exosomes are one of life's shredding machines.

to splice them together, in exactly the right order, to get the code for the machine part/protein it needs?

Good questions. Accessing the correct part of DNA is called "gene expression." Gene expression is complex and not completely understood. It seems to work in part by chemical clues; in some cases, a specialized machine comes along and chemically "tags" the appropriate section of DNA for copying. How does your cell know exactly what stretch to read when it needs to build a particular protein—exactly where to start, and where to stop, in 6,000 miles of rope-sized DNA? We don't fully know yet, but we can marvel at this ability.

I'll give you a clue. The DNA that does not contain a gene, the 98.5 percent of your DNA that is "noncoding," seems to be where this kind of information is kept. That was the stunning announcement in September 2012 by 450 scientists working together on the "ENCODE Project." We'll come back to the ENCODE Project in the next chapter, as we look at the myth of "junk DNA."

Another question I had, that I didn't find addressed anywhere in the literature, is how all of this information, such as your 3.2 billion letters of DNA, could possibly be input correctly? When I typed a single letter wrong while punching any of hundreds of computer cards, the program almost always didn't work. I think people today tend to forget this, because human technology has advanced to include error correction. Your smartphone will generally correct your text, although it often guesses wrong. Word processing technology has this capability. The human brain has this ability, you can read words with numerous spelling and grammatical mistakes and still understand the intent. (I ken gwhereentee dat es twue.) But these are examples of advanced technology and intelligence; error correction processes do not arise by accident. So either our bodies read DNA with error correction technology, or the DNA in our bodies has been fantastically assembled in almost perfect order, or both. Whichever is true, it is scientific evidence of wonder.

Allow me to share more of the technology of life.

Molecular Factories

So now the proper section of DNA—the correct gene—has been opened up and the information transcribed into messenger RNA. In eukaryotic cells (cells with a nucleus) the messenger RNA may have been correctly spliced. In eukaryotic cells, the messenger RNA has been transported outside the nucleus. The cell is finally ready to build a biological machine part, a protein. "Proteins are by far the most structurally complex and functionally sophisticated molecules known."[14]

I find this next step mesmerizing. It is highly logical, highly organized, and screams of design. When I first read how it works, I was puzzled, because it seemed that anyone who knew about this process should believe in God.

Your cells contain factories called "ribosomes" that read the code in messenger RNA and construct the proteins necessary for life. The ribosome is an awesome assembly plant, an ancient molecular juggernaut made up of two main chains of RNA and more than fifty different proteins precisely assembled.[15] All life has ribosomes that work roughly the same way. A human ribosome can read the code in DNA from a bacteria to assemble the proteins a bacteria needs, and a bacterial ribosome can read the code from humans to assemble the proteins humans need.

The Genetic Code

It took over ten years to decipher DNA's four-letter code. We now know the ribosome reads the code three letters at a time. With four letters, there are $4 \times 4 \times 4 = 64$ three-letter combinations. Each tells the ribosome to start, to add a particular amino acid to the protein it is building, or to stop.

The ribosome begins by reading the first three letters of messenger RNA. It "knows," using the logic of the "genetic code" shown in the figure on the next page, which of the twenty possible amino acids that particular three-letter code refers to.

The ribosome selects and holds in place the correct amino acid to start building the protein. It then reads the next three letters. It uses the genetic code to know which of the twenty possible amino acids that second three-letter code refers to. It selects that second amino acid and connects it to the first. It reads the next three letters, selects the correct amino acid, and connects that third amino acid to the second one. And so on. As messenger RNA rolls through the molecular factory, the ribosome reads the code three letters at a time, pulls out the correct amino acid (of course, the ribosome has a supply of amino acids for building proteins), and adds it to the

Genetic Code

		Second Base				
		U	**C**	**A**	**G**	
First Base	**U**	UUU ⎤ Phe UUC ⎦ UUA ⎤ Leu UUG ⎦	UCU ⎤ UCC ⎤ Ser UCA ⎦ UCG ⎦	UAU ⎤ Tyr UAC ⎦ UAA Stop UAG Stop	UGU ⎤ Cys UGC ⎦ UGA Stop UGG Trp	U C A G
	C	CUU ⎤ CUC ⎤ Leu CUA ⎦ CUG ⎦	CCU ⎤ CCC ⎤ Pro CCA ⎦ CCG ⎦	CAU ⎤ His CAC ⎦ CAA ⎤ Gln CAG ⎦	CGU ⎤ CGC ⎤ Arg CGA ⎦ CGG ⎦	U C A G
	A	AUU ⎤ AUC ⎤ Ile AUA ⎦ AUG Met Start	ACU ⎤ ACC ⎤ Thr ACA ⎦ ACG ⎦	AAU ⎤ Asn AAC ⎦ AAA ⎤ Lys AAG ⎦	AGU ⎤ Ser AGC ⎦ AGA ⎤ Arg AGG ⎦	U C A G
	G	GUU ⎤ GUC ⎤ Val GUA ⎦ GUG ⎦	GCU ⎤ GCC ⎤ Ala GCA ⎦ GCG ⎦	GAU ⎤ Asp GAC ⎦ GAA ⎤ Glu GAG ⎦	GGU ⎤ GGC ⎤ Gly GGA ⎦ GGG ⎦	U C A G

(Third Base)

string, sort of like you might build a necklace by snapping together plastic beads or build a sentence by typing letters and spaces.

It would take a separate book to fully describe the wonders of the genetic code. I'll briefly note three. First, the twenty amino acids that form the "alphabet" of life, that are assembled in various ways to form proteins, appear to have been selected very carefully. A 2011 study compared life's "standard alphabet" of twenty amino acids for "size, charge, and hydrophobicity [aversion to water] with equivalent values calculated for a sample of 1 million alternative sets (each also comprising 20 members) drawn randomly from the pool of 50 plausible prebiotic candidates." The two NASA astrobiology scientists who conducted the study found "the standard alphabet exhibits better coverage (i.e., greater breadth and greater evenness) than any random set." (The standard alphabet was better than *any* of 1 million randomly chosen possible sets of twenty amino acids.) They conclude, "[T]he standard set of 20 amino acids represents the possible spectra of size, charge, and hydrophobicity more broadly and more evenly than can be explained by chance alone."[16] Second, the genetic code, in the way it "maps" the 64 three-letter DNA combinations to life's alphabet of twenty amino acids, is a "marvel"[17] of error correction and error minimization technology. A complex statistical study found it "highly optimal" because it significantly outperformed 1 million randomly chosen alternate codes.[18] Third, there is no known way the genetic code could have "evolved." Here's again Eugene Koonin, senior investigator at the National Center for Biotechnology Information, National Library of Medicine, National Institutes of Health, and a recognized expert in the field of evolutionary and computational biology, with coauthor Artem Novozhilov in 2009:

> Summarizing the state of the art in the study of the code evolution, we cannot escape considerable skepticism. It seems that the two-pronged fundamental question: "why is the genetic code the way it is and how did it come to be?," that was asked over 50 years ago, at the dawn of molecular biology, might remain pertinent even in another 50 years. Our consolation is that we cannot think of a more fundamental problem in biology.[19]

To me, the genetic code is a miracle. In life's standard alphabet of twenty amino acids, in the genetic code's quality of minimizing error, and in the difficulty of even remotely conceiving of how one genetic code could possible evolve into another, I see the hand of a master designer. I cannot believe that the genetic code, with all of the molecular technology to read it and assemble proteins, arose by a chance-based process. It appears that the genetic code was created 3.5 billion years ago.

Building Proteins

Human proteins vary greatly in length, but a "typical" human protein has about 430 amino acids. An amino acid is a molecule formed by an "amino group" (a nitrogen atom plus hydrogen atoms), plus a "carboxyl group," which consists of carbon, oxygen, and hydrogen atoms, plus a "side train" of atoms. The 20 amino acids used to make proteins all have a central carbon atom to which is attached the ammonia part, the carboxyl group, and the side chain. A carbon atom can share electrons with, and thus bond strongly with, four different atoms. The central carbon atom in each amino group in a protein, the a-carbon atom, connects the amino group, the carboxyl group, the side chain, and a hydrogen atom. These four connections are spaced evenly in three-dimensional space to form a tetrahedron.

Because it is a three-dimensional object, each amino acid can be formed in one of two ways, which are mirror images, or optical isomers, of each other. In other words, if you take the amino acid and put the lone hydrogen atom connected to the a-carbon atom at the top, the other three connections can be made in two different ways that are mirror images of each other. There is no known law of physics that dictates a preference for one of these mirror-image shapes over the other. Yet almost all of the amino acids in living organisms are left-handed. In other words, with the lone hydrogen atom at the top, the direction from the amino group to the carboxyl group almost always forms a left-handed molecule.[20] This fact also made me stop and take a breath. Is it not perhaps evidence of design?

The side chains of the twenty amino acids that are used to build proteins have different chemical properties. Some of these twenty side chains have electric charges, some are uncharged, some are attracted to water molecules, some repel water molecules, and so on. These different chemical properties drive the technology of life; they are why proteins become molecular machine parts.

The long, constructed chains of amino acids fold, or are folded, into a functional machine part/protein. Some estimate that less than one in a billion linked chains of amino acids will fold into a stable, three-dimensional shape, yet the vast majority of proteins fold into a stable shape.[21] How a string of amino acids folds into a protein is not completely understood. In many cases, it appears that the inherent electrical and chemical nature of the structure causes it to fold in the proper way. In some cases, other proteins assist.

Life has a lot of amazing technology, functional nanotechnology, technology built atom by atom. The closer you look at it, the harder it is to believe it all arose by chance. Let me show you more.

Information Compression Technology

Your body has technology to store and compress DNA. The first step is to wind your DNA on spools with grooves designed to hold it. Each of these spools, called "nucleosomes," holds a little less than two wraps (1.65 turns, or about 147 letters) of DNA. Nucleosomes are built from eight specialized interconnecting proteins—called "histones"—that hold your DNA in their grooves and contain "tails"—special chemical groups that can be triggered to instruct the nucleosome to release or unwind your DNA so it can be accessed. Almost every cell in your body has about 30 million nucleosomes.* Each nucleosome has about 147 letters of DNA wrapped around it. There is "linker DNA" between these spools, so there is a nucleosome for about every 200 letters of DNA. The nucleosomes are precisely contoured to hold your DNA tightly; in each, 142 "hydrogen bonds" (where a hydrogen atom shares an electron with an adjoining atom) and other chemical reactions hold on to the backbone of your DNA.†

The next step is to arrange nucleosomes to form a "chromatin fiber." This fiber has repeating "zigzag" arrays of nucleosomes. Those tails on the histone proteins in the nucleosomes help the nucleosomes arrange themselves to create the chromatin fiber, and other proteins assist.

Your DNA is most compact when your cell is ready to reproduce. At this stage, it has somehow been compacted another 50 times, so that it is almost 10,000 times as compact as a single strand of DNA.‡

If you came across systems half as complex in any other setting, you would "know" they had been designed.

Shredding Technology

Unused or waste RNA is dangerous. It is shredded by disposal machines called "exosomes," which exist in all life. There is a hollow barrel made up of nine proteins. RNA is channeled down the barrel to reach a tenth protein, a "catalytic subunit," which shreds the RNA. "Cells lacking any of the ten proteins do not survive and this shows that not only the catalytic subunit but also the entire barrel is critical for the function of

---

* Red blood cells are the primary exception.

† For the most part, these hydrogen bonds and other chemical reactions do not depend on the interior chemical groups—DNA letters—within the double helix, so that again any letter can appear at any location. However, some sequences of interior letters do result in tighter bonds, and nucleosomes can position themselves—"roll" along the DNA—as necessary for the cell to operate.

‡ When your DNA is most compact, it cannot be "accessed" or read to build proteins. It is believed that, in an ordinary cell, there are regions of DNA—called "heterochromatin"—that are always this compact and cannot be read. Only in special "stem" or germ cells is this portion of DNA accessed.

the exosome."[22] Exosomes appear to be "irreducibly complex," a concept we will look at in the next chapter.

Replication Technology

Replicating 3.2 billion letters of DNA in one of your cells is quite a feat. The replication technology is fundamentally similar in bacteria and complex life, and in each it is accurate to about one letter in a billion.[23] If the error rate were ten times greater—say, one mistake in every 100 million DNA letters copied—"evolution would probably have stopped at an organism less complex than a fruit fly."[24]

Molecular biologist James Shapiro of the University of Chicago is in awe of life's replication technology:

> It is essential for scientists to keep in mind the astonishing reliability and complexity of living cells. Even the smallest cells contain millions of different molecules combined into an integrated set of densely packed and continuously changing macromolecular structures. Depending upon the energy source and other circumstances, these indescribably complex entities can reproduce themselves with great reliability at times as short as 10–20 minutes.
>
> Each reproductive cell cycle involves literally hundreds of millions of biochemical and biomechanical events. We must recognize that cells possess a cybernetic capacity beyond our ability to imitate. Therefore, it should not surprise us when we discover extremely dense and interconnected control architectures at all levels.[25]

When one of your cells is ready to reproduce, your DNA is opened up at various points, called "replication forks." The double helix is separated into two strands. Each of these strands contains the entire code. Your body takes a single strand of DNA and uses it as a template to build a new double helix. It creates an opposite strand of DNA with the correct opposing chemical groups. An A on one strand is always paired with a T on the new opposing strand, and a C is paired with a G. The "replication machine" that does this works at 1,000 letters of DNA a second. It is efficient. It reads a single strand of your DNA and builds opposite it a new DNA subunit with the precise correct interior groups. This requires a supply of DNA subunits that the machine can access. This replication machine makes only about one mistake every 100,000 DNA letters. That is incredible accuracy; if you could type with that accuracy, you could type a midsized novel and expect to hit only a handful of wrong keys.*

I found this description of how it works for *E. coli* helpful. If the DNA strand were magnified to have a diameter of one meter, the replication machinery would be the size of a Federal Express truck, move down the line at 375 miles per hour, and make a mistake about once every 100 miles.[26] Keep in mind this machinery synthesizes both new chains simultaneously.

When human DNA is replicated, more than one hundred of these machines work simultaneously on each of the forty-six DNA molecules (chromosomes). It is not clear how these one-hundred-plus replication machines are able to coordinate their work.

Other machinery, other nanotechnology, comes behind, "proofreads" the assembled doubled helix, and corrects about 9,999 out of every 10,000 initial errors. The technologies and machines to do this differ among organisms but achieve similar results.† It is astonishing that the final product of

---

* The replication process is again more intricate than you might suspect. The DNA strands, when separated, "point" in different directions. The replication machine, however, can continuously copy DNA in only one direction. This works fine for one of the strands, called the "leading" strand—the replication machine rolls merrily along creating a new double helix from a single strand of DNA. However, for the other strand, the "lagging" strand, the replication machine periodically "backfills" a new strand of DNA. It waits until 100 to 200 letters of DNA are opened on the lagging strand and then backfills a new opposing backbone and chemical groups for these 100 to 200 letters. These fragments of DNA—called "Okazaki fragments" after their discoverer—are ultimately connected to become a smooth opposing backbone. Somehow, when the molecular machine doing this reaches the end of the single strand and touches a portion of DNA to which an opposing backbone has already been added, it knows to stop.

† The first proofreading mechanism is built into the replication machine. Before a new DNA subunit or letter is added—again, this subunit has both a backbone piece and an A, C, T, or G amino acid—the replication machine checks whether the last replication was done correctly. If not, it attempts to fix it before continuing. This process corrects about 99 out of every 100 initial errors. The second proofreading step is a separate machine. It looks for mismatched pairs—such as an A matched with a C and not with a T—and corrects the mismatch. This mismatch repair machine has a chemical technique for distinguishing the new DNA strand from the original DNA strand. If it did not, it would not know which of the mismatched letters was in error. These "mismatch" repair machines work in three steps: they locate a mismatch, cut out the mismatched DNA subunit, or nucleotide, and insert a new subunit to correct the error. This second process corrects about 99 out of every 100 errors. These three steps then—initial strand building correct to one letter in 100,000, immediate checking to repair 99 out of 100 mistakes, and a separate "mismatch repair" molecular machine that corrects about 99 out of every 100 remaining mistakes—are able to reproduce your DNA with an error rate of 1 letter in 1 billion. This accuracy makes complex life possible.

DNA replication in both bacteria and human beings differs from the original on average only by one letter in 1 billion—1 in 1,000,000,000.

It appears that the technology and machinery for replicating DNA in cells with a nucleus has not changed significantly in hundreds of millions of years. In the language of biology, it has been "conserved" to a great extent since cells with a nucleus came into existence 2.7 billion years ago.[27]

Repair Technology

DNA has to be maintained. Spontaneous chemical reactions alone change about 5,000 letters of DNA each day in each cell.[28] If left untreated, these changes would create an unacceptable rate of mutation.

Cells have different types of incredible repair machines that fix more than 99.9 percent of all DNA damage.[29] Many human diseases—colon cancer; skin cancer; leukemia; breast, ovarian, and prostate cancers; stunted growth; and others—have been linked to problems with DNA repair.

You do not need a PhD in molecular biology to appreciate the wonder of the technology of life. I realize that, no matter how many incredible machines and processes are described to them, many people will not be able to see outside of the prevailing paradigm. Darwinists claim all this technology, all of the incredible machines and processes in our bodies and other forms of life, arose from accidental mutations and natural selection. I see the hand of a master designer. To me, the discoveries of the technology of life are the fourth wonder of modern science.

Return to Probability

Consider the protein, the biological machine part, called "histone H4." It's one of four different proteins used in pairs to build the spools—the nucleosomes—around which the DNA of eukaryotic cells is wrapped. Each of your cells has about 30 million nucleosomes, and therefore about 60 million copies of the protein known as histone H4.

Histone H4 is relatively short: 102 amino acids folded exactly right. In biological terms, histone H4 is "highly conserved." Perhaps because of its key role in holding and releasing DNA, histone H4 has changed or evolved little in hundreds of millions of years. Of the 102 amino acids, 100 are the same in a pea and a cow. It appears that the particular order of these 100 amino acids is essential for this vital piece of cellular machinery to function properly. It has been proven that almost all changes in the amino acid sequences of histones are lethal or cause serious abnormalities.[30]

It is not hard to estimate the odds that accidental processes, even given an ample supply of the 20 amino acids necessary to build proteins, would accidentally create a single copy of histone H4. To simplify, what are the odds of getting, by dumb blind luck, a specified sequence of 100 amino acids? When we are done linking 100 amino acids, the odds that we did it correctly are 1 in 20 multiplied by itself 100 times, or one in 20^{100}.

Now 20^{100} is a monster number. It is about the same size as one in 10^{130}, one in a number with 130 zeros. We're down 130 levels in our magical building of chapter 10, within sight of Dembski's universal probability limit of one in 10^{150}, and this is just one protein.

Consider a ball extending 25 trillion miles from Earth in all directions, reaching the nearest star, packed with marbles one-half inch in diameter. Imagine hidden somewhere in that immense ball is a special, "marked" marble that you need to find. There are about 5 times 10^{55} marbles in that ball, and so your chances of blindly picking that special marble are trillions and trillions of times more likely than getting a specified sequence of 100 amino acids.

And even if the odds of 1 in 10^{130} were overcome, and a copy of histone H4 was created by accident, you have only begun. To build a single nucleosome, you need to start with two copies of histone H4 and six other proteins, as well as the information to put these parts together correctly. To use nucleosomes, you need technology to wind and unwind DNA off the spools.

Nanotechnology is technology built at the atomic level. Functional nanotechnology, such as technology for coiling and accessing DNA on nucleosomes, does not arise by accident. It doesn't matter how many Earth-sized planets are out there, or how many billions of years have elapsed. You still have no plausible chance of creating functional nanotechnology. That is a mathematical truth, not a religious statement. You can't get functional biological nanotechnology—both the organic machines and the technology to operate them—by any mechanism involving only blind chance. Any person who thinks otherwise does so based on materialism/Scientism/Atheist beliefs, not because of science.

We have now counted to God through four wonders of modern science—creation, the fine-tuning of the universe, the origin of life, and the technology of life. We will count to five as we look at the puzzles of macroevolution, the creation of radically new forms of life.

CHAPTER 12

Puzzles of Macroevolution

How did life get so complex?

If it could be demonstrated that any complex organ existed which could not possibly have been formed by numerous, successive, slight modifications, my theory would absolutely break down.

CHARLES DARWIN

THE TERM "macroevolution" refers to major changes in species. To me, the puzzles of macroevolution, and the growing mountain of evidence against neo-Darwinian theory, are the fifth wonder of modern science, the fifth of seven in our count to God. The last chapter explored some of the wonder in the technology of life, such as DNA. This chapter explores some of the wonder in the creation of wholly new species, including human beings.

As I dug deeper into the science of belief, I was forced to confront the subject of evolution. I was hesitant to do that. I majored in math and physics, not biology, and I knew I had a lot to learn. I also knew I was entering a field of enormous controversy. Perhaps nowhere is the animosity between Belief and Scientism greater than in the debate about evolution.

I knew evolution is taught just about everywhere. Clearly, organisms have developed and changed over time. Modern DNA analysis suggests connections between species. Apes look sort of like people, don't they? And the conflict seemed unnecessary. Couldn't God have used evolution to create human beings?

Some writers skip the evolution debate altogether. In *Nonsense of a High Order—The Confused and Illusory World of the Atheist,* Rabbi Moshe Averick calls it "the ideological equivalent of the Battle of Stalingrad" and refers to Darwinian evolution as "the Maginot line of atheism."[1] His biology discussion focuses on the origin of life, and I think rightly so, for the origin of the information necessary to build life is the Achilles' heel of neo-Darwinian theory. But I knew enough to tantalize, and my curiosity would not let me look away. So my journey took me to evolution, and this chapter is about the puzzles and the wonder that I found.

We'll start by defining "evolution," and we'll see that the only two explanations for macroevolution are neo-Darwinian theory and design. We'll see that the predictions of neo-Darwinian theory do not match the facts, such as in the fossil record. We'll look at theories die-hard Darwinists use to explain the fossil gaps. Then we'll look at three other areas where the predictions of neo-Darwinian theory do not match the facts: "junk" DNA, irreducible complexity, and orphan genes. Finally, we'll marvel over the creation of human beings.

Two Explanations

What is "evolution"? The first meaning, according to one online dictionary, is "any process of formation or change." Everyone I know agrees that life changes. Museums display skeletons of massive dinosaurs, and they sure don't walk around anymore. Many species are now extinct; you can find drawings of the Dodo bird, which was hunted to extinction in the late 1600s. You often read about other species under pressure to survive.

The real controversy is over what causes "macroevolution." What creates radically different organisms with radically different biological systems and body parts? What explains the variety of life, the differences between an amoeba, an ant, a catfish, a wolf, a hummingbird, and a human being?

Organisms sometimes do change, or "evolve" in the broad sense of the word, for better or for worse, over time. It is clear, and accepted scientific fact, that changes that increase an organism's ability to survive and reproduce are more likely to be carried on to future generations. This is called natural selection. Accidental, random, haphazard changes in DNA do occur, and cause mutations. Most, probably far more than 99.99 percent, are harmful or have no discernible effect. However, some may accidentally confer a slight advantage, and make that organism more likely to survive and more likely to have offspring that survive. It makes sense that a gene with a reproductive advantage is likely, given enough time, to slowly spread throughout a population.

But does natural selection have the ability, is it a strong enough force, to create wholly new species? That is the debate, and it was a debate right from the beginning. The cofounder of the theory, Alfred Russel Wallace, disavowed natural selection in 1869, in favor of what Wallace called "intelligent evolution."[2] In the 1800s, natural selection was often referred to as "the Darwin and Wallace theory."

Well before genetics was a science, Charles Darwin and others noted that organisms appeared to adapt to changes in their environment. The common perception is that Darwin studied changes in the beaks of finches in the Galapagos Islands off South America, and he certainly did travel to the Galapagos, although recent scholarship suggests the finch story was contrived a century later.[3] In any event, about thirty years after his visit to the Galapagos, and after correspondence with Alfred Russel Wallace, Darwin published his theory of natural selection. Darwin and Wallace were not the first to notice that organisms were suited to their environment and that the healthiest, most fit individuals tended to have more offspring. That was suggested years earlier by others, such as Edward Blyth. But Blyth thought natural selection simply maintained the stability of species by weeding out "unfit" members; Blyth did not think natural selection could create new forms of life. Charles Darwin suggested it could. In his famous book *On the Origin of Species,* Darwin suggested that natural selection alone could create the tremendous variety of life we observe today. According to the college textbook *Molecular Biology of the Cell,* "[I]t is estimated that there are more than 10 million—perhaps 100 million—living species on Earth today."[4] Darwin claimed that the gradual process of accidental mutation and natural selection is solely responsible for this variety.

This belief in accidental mutation and natural selection alone is the central tenet of the modern version of Darwin's theory, which again is called neo-Darwinism or neo-Darwinian theory. Neo-Darwinists proclaim "evolution is a fact," by which they mean that Darwin's theory of what causes macroevolution is a fact. They are wrong; change is a fact, what causes change is a theory. Neo-Darwinian theory is under increasing attack, from Atheists as well as from religious persons, because its predictions do not match the facts and because it does not provide a mathematical model for the creation of radically different biological systems and body parts. There is general agreement that natural selection can gradually alter species, but many conclude it does not have the ability to create new forms of life.

After that 2005 debate in Boston on intelligent design (see chapter 6), I learned that *hundreds* of scientists, yes *hundreds,* do not believe that accidental mutations and natural selection alone could have resulted in many of the complex biological systems in living organisms. One website[*] lists over 850 scientists (as of August 2013) who agree with the following statement:

> We are skeptical of claims for the ability of random mutation and natural selection to account for the complexity of life. Careful examination of the evidence for Darwinian theory should be encouraged.

I learned the fossil record does not contain the myriad transitional forms of evolving organisms predicted by Charles Darwin. I learned that new biological systems typically require complex new biological parts (proteins) and detailed new instructions (DNA coding) on how to make the parts and put the parts together, each of which is almost mathematically impossible to arise by accident. I learned that many of these systems are "irreducibly complex"; they do not work unless all the parts are present and assembled in the right way. I learned that accidental mutations generally destroy information and are not likely to produce a new type of life. I learned that what causes macroevolution—the creation of radically new organisms with radically new biological systems and body parts—is very much the subject of legitimate scientific debate.

---

[*] dissentfromdarwin.org. This is a list of PhD scientists. Because scientists who put their name on this list are frequently persecuted, I believe there are thousands of other scientists who are skeptical of neo-Darwinian theory but who are not willing to authorize the inclusion of their name on this list.

The more than 850 scientists who question the neo-Darwinian theory of macroevolution do not dispute that species change over time. They do not dispute that natural selection causes species to adapt, over time, to changes in their environment. What they do dispute is that accidental mutations and natural selection alone could have created the enormous diversity of life, and they particularly question whether accidental mutations and natural selection can create radically new biological systems and body parts. In other words, it has been demonstrated that natural selection can result in some birds having slightly bigger beaks than others, but it is not documented, and not even reasonable, some scientists claim, to think that natural selection alone created complex systems such as the eye, the backbone, or the human brain. To put it another way, natural selection may adjust the size of the finch beak, but can it create the finch? "The problem is not the survival of the fittest, but the arrival of the fittest."*

Many of these 850 scientists are Atheists. In 2012 noted Atheist philosopher Thomas Nagel released a new book, *Mind and Cosmos: Why the Materialist Neo-Darwinian Conception of Nature Is Almost Certainly False.* Nagel states that "proponents of design deserve our gratitude" for challenging neo-Darwinian theory. He admits honestly that his Atheism prevents him from embracing intelligent design.

So, as we asked at the beginning of this chapter, "How did life get so complex?" At this time there are only two competing explanations. One is neo-Darwinian theory, accidental mutations and natural selection. The other is design. It's a stark choice, and the Atheists such as Thomas Nagel who recognize the flaws in neo-Darwinian theory seek a third choice. But they have nothing to propose. For now at least, it's Darwin or design. If you want to explain how life got so complex, they are the only games in town.

Neo-Darwinian theory has a clear appeal. It is a straightforward explanation, based on observation, for the enormous diversity of life. The fourth floor of the American Museum of Natural History in New York City has an incredible exhibit that is, in a way, a shrine to evolution. It takes you from the first creatures with a brain case and the beginning of a backbone (vertebrates) around 500 million years ago, to the development of fish jaws 430 million years ago, to the creation of four-legged species (tetrapods) 380 million years ago, to the creation of amniotes 290 million years

---

* This quote is from Jerry Bergman.

Neo-Darwinian Predictions versus Facts

Subject	Neo-Darwinian Prediction	Fact
Fossil record of when major new species appear	Step-by-step evolution of major new species	Periods of "explosions" when multiple new body plans and species suddenly appear
Development of major new species over time	As the diversity of life accelerates, major new species more likely to appear	Greatest changes occurred from 541 to 516 million years ago
Fossil record of transition organisms	Overwhelming evidence of transition organisms leading to the creation of major new species	Very few transition organisms between major groups
Stretches of Noncoding DNA	Accidental mutations create large sections of useless, noncoding "junk" DNA	DNA contains multiple layers of information; noncoding sequences control gene expression and serve other critical functions
Designs of Living Systems	Generally, new systems are created step-by-step from accidental mutations, each of which by itself confers a benefit to the advantage of the organism	Generally, living systems are irreducibly complex; necessary proteins confer no advantage unless all present and properly assembled
Orphan Genes	All genes are related; new genes and their associated proteins develop gradually from genes/proteins in existing organisms	All organisms have "orphan" genes with no known relatives in other organisms; these genes often make the organism unique[5]

ago with a watertight membrane for embryos, and so on. As you walk through the halls with exhibit after exhibit, you get a clear sense of life evolving over millions of years into the forms that exist today. If you just look at the fabulous fossils, it is reasonable to imagine all of these changes occurred as a result of gradual, incremental changes, in the manner proposed by Charles Darwin.

But if you put neo-Darwinian theory under the microscope of science, and probe it fully with human observation, experimentation, and logic, you are likely to conclude that, when it comes to macroevolution at least, something is quite wrong. As the table on the preceding page shows, the predictions of neo-Darwinian theory do not match the facts.

The Fossil Record

When it comes to evolution, there is fact and there is theory. Let's start with the facts, and then see which theory best fits the facts. And as for facts, let's start with the fossil record.

About 150 years ago, when *On the Origin of Species* was first published, the fossil record was mostly a mystery. Today we know a lot. Although there are surely new discoveries to be made, the overall picture has come into fairly clear focus, as revealed by the research of thousands of scientists.[*]

Rocks around the world tell the same general story. Fossilized bacteria are found in rocks 3.465 billion years old.[6] Life developed fairly slowly after that for close to 3 billion years. Around 570 million years ago, some wholly new creatures appeared, but, with the primary exception of sponges, those organisms became extinct and their body plans "bear no clear relationship" to any later organism.[7]

Then, 541 million years ago, life exploded. Rocks around the world, especially the Burgess Shale in Canada and the Maotianshun rocks of China, tell a consistent and amazing story. During perhaps 25 million years, life on Earth exploded from mostly single-celled organisms to multiple-celled organisms with complex bodily structures. This burst of

---

[*] "If scaled to the . . . taxonomic level of the family, the past 540 million years of the fossil record provide uniformly good documentation of the life of the past." M. J. Benson, M. A. Wills, and R. Hitchin, "Quality of the Fossil Record Through Time," *Nature* 403 (February 3, 2000): 534–36. See also *Darwin's Doubt,* pp. 69–71, explaining why statistical paleontology strongly suggests that "in many respects our view of the history of biological diversity is mature."

creativity is called the "Cambrian explosion." It is fact, not theory, and it is directly contrary to neo-Darwinian models of gradual change. Darwin said: "If numerous species, belonging to the same genera or families, have really started into life all at once, the fact would be fatal to the theory of descent with slow modification through natural selection."[8]

The "animal kingdom" is divided into 33 "phyla," at least according to one group of experts.[9] A "phylum" is a high level of classifying animals according to their general body plan. If we were looking at machines for transportation, cars, boats, and planes would be in different phyla. In the animal kingdom, spiders and barnacles are in the same phylum (they both have jointed legs and exoskeletons at some stage in their life cycle), but earthworms and tapeworms, although much closer in appearance, are in different phyla.* We humans are in the chordate phylum, along with all other mammals, fish, birds, reptiles, and amphibians. We share a notochord, a hollow dorsal nerve cord, pharyngeal slits, an endostyle, and a postanal tail for at least some part of our life cycles. For a full discussion of these terms, go ask your mother.

According to the above group of experts, of today's 33 animal phyla, 23 came into existence during the Cambrian explosion, at the beginning of the Cambrian era.[10] Two predate the Cambrian, two came later in the Cambrian era, and six arose after the Cambrian era. So 70 percent of all body plans of all of today's animals were created in a relatively short period of time beginning 541 million years ago. Most of this explosion in animal life may have occurred in as little as 6 million years. According to Stephen Meyer in his 2013 bestseller, *Darwin's Doubt*:

> An analysis by MIT geochronologist Samuel Bowring has shown that the main pulse of Cambrian morphological innovation occurred in a sedimentary sequence spanning no more than 6 million years. Yet during this time representatives of at least sixteen completely novel phyla and about thirty classes first appeared in the rock record. In a more recent paper using a slightly different dating scheme, Douglas Erwin and colleagues similarly show that thirteen new phyla appear in a roughly 6-million-year window.[11]

---

* Usually it is pretty clear that organisms from different phyla are different from one another. A clam is not like a shrimp or a fish or a sea star. The worm phyla are an exception. Earthworms, nematodes, and tapeworms, for example, look superficially similar, but they are in different phyla because of the details of their anatomy not apparent to an untrained eye.

That is not what a Darwinist would predict. Neo-Darwinian theory predicts the opposite. Neo-Darwinian theory predicts that, over time, as the number of species on Earth expanded and the cumulative gene pool grew, one would expect accidental mutations to multiply and create more and more major phyla—animal body plans. As life diversified, the creation of major new species should accelerate. Yet the fossil record reveals that the most stunning changes in evolution occurred in a very short period. If the history of life on Earth were compressed into twenty-four hours, 6 million years would be less than three minutes.

It is not just the sudden and tremendous variety of new life that contradicts neo-Darwinian theory. A second contradiction is the "top-down" nature of the fossil record, where we first see major new body plans and very different forms of life, and then variations on those themes. Neo-Darwinian theory predicts the opposite; it predicts minor variation leading slowly to greater changes.

The Cambrian explosion has been called "biology's big bang." As one scientist puts it, "[N]ew bodily designs appear in the fossil record as Athena did from the head of Zeus—full blown and ready to go."[12] Charles Darwin was aware of this relatively sudden explosion in the fossil record, but he believed the record was incomplete, and that time and further research would fill in the gaps. That has not happened. He knew that a lack of transitional forms would be "the most obvious and serious objection which can be urged against the theory."

The fossil record has other "explosions." "The Siluro-Devonian primary radiation of land biotas [plant and animal life on land] is the terrestrial equivalent of the much-debated Cambrian explosion of marine faunas."[13] As for the sudden appearance of fish species, "[T]his is one count in the creationists' charge that can only evoke in unison from paleontologists a plea of nolo contendere [no contest]."[14] In the "Early Tertiary" explosion, "many bird and mammal groups appear in a short time period lacking immediately recognizable ancestral forms."[15]

A third way the fossil record contradicts neo-Darwinian theory is that there are few, if any, transitional organisms. According to Darwin:

> The number of intermediate varieties, which have formerly existed, [must] be truly enormous.[16]

I think today all scientists, even die-hard Darwinists, admit the fossil record is contrary in this third way to Darwin's expectations. "The known fossil record fails to document a single example of phyletic evolution accomplishing a major morphologic transition and hence offers no evidence that the gradualistic model can be valid."[17] Darwinist paleontologist Niles Eldredge concedes "[n]o one has found any 'in-between' creatures: the fossil evidence has failed to turn up any 'missing links,' and many scientists now share a growing conviction that these transitional forms never existed."[18] His coauthor, staunch Darwin defender Stephen Jay Gould, admits: "The absence of fossil evidence for intermediary stages between major transitions in organic design, indeed our inability, even in our imagination, to construct functional intermediates in many cases, has been a persistent and nagging problem for gradualistic accounts of evolution."[19]

New York's American Museum of Natural History hasn't gotten this message. In one of its rooms of fabulous fossils, on the wall behind a superb skeleton of *Tyrannosaurus rex,* is archaeopteryx, a bird fossil with reptilelike teeth, and a statement that this single creature validates Darwin's theories. It is not clear how archaeopteryx fits in the history of life. Modern birds are probably not descended from it, and it appears in the fossil record tens of millions of years before the dinosaurs it supposedly is descended from,[20] but that is not what I find most objectionable. What I find most objectionable is that, even if archaeopteryx was a true transitional creature, it has little company in the fossil record, and it is plain error to claim that a single odd creature validates neo-Darwinian theory. The "truly enormous" number of "intermediate varieties" predicted by Charles Darwin does not exist.

New York's Museum of Natural History should not get a pass from intellectual honesty. Stephen Jay Gould said in 1977, "All paleontologists know that the fossil record contains precious little in the way of intermediate forms; transitions between the major groups are characteristically abrupt."[21] He called it the "trade secret" of paleontology. It is sad that the American Museum of Natural History in New York makes false claims to prop up Darwin's theory, but not surprising because the reigning paradigm of our society does not permit honest intellectual debate about the possibility of design.

So how do Gould and Eldredge attempt to prop up Darwin's theory, to explain away these three contradictions to what Charles Darwin predicted, to plug the gaps in the fossil record? In 1972 they proposed "punctuated equilibrium," where most evolution takes place rapidly in small populations that do not leave a fossil record. But why and how? How does that result from natural selection, from the gradual accumulation of accidental changes? Punctuated equilibrium offers no explanation for why rapid change would ever occur in a short period of time in a small population. It merely assumes that, because that's what the fossil record suggests, it must have resulted from natural selection and accidental mutations.

That's the logical gap of neo-Darwinian theory. Whatever you observe, you just say "evolution caused it."

IT EVOLVED   (C) 2008 B.A. MILLER

You know, Darwin sure has made biology simple.

It makes my head swim. It's so much easier to say "it evolved."

Yah, saying "it evolved" is so much easier for reputable biologists to agree on.

I know. Can you imagine how much extra work we would have to do to prove where each mechanism evolved from? We'd never publish anything.

Isn't that the truth, plus saying "it evolved" already has consensus and goes right through peer review.

You don't have to explain, and you don't have to give a mathematical analysis of the likelihood of a beneficial mutation or of multiple favorable genes spreading through a population (Fred Hoyle did the math in 1987 and emphatically concluded it is impossible).[22] You explain the survival of an organism because of its fitness, and its fitness because of its survival, and you disregard that the logic is obviously circular. As Darwin doubter David Berlinski says, "This is not a parody of evolutionary thinking; it is evolutionary thinking."[23] The neo-Darwinian theory of evolution cannot be falsified because, whatever you observe, "evolution caused it."

Where does the information come from? Accidental mutations generally destroy genetic information. Suppose you take a page of text, and then start "mutating" the letters and spaces—you randomly change some letters/spaces into others, take letters/spaces out in random places, and add new letters/spaces in random places. How likely is it that you will transform the text into something readable and coherent? Not likely. Sure, it can be done, and if you as an intelligent agent purposefully "mutate" letters and spaces, you can make it work. But without intelligence, without design, without purpose, the odds are vanishingly small. In chapter 10, we saw how impossible, how mind-blowingly unlikely, it is that monkeys might ever type at random even a short snippet of Shakespeare.

In the figure below are two similar proteins from *E. coli*, a bacterium that lives in human guts (fortunately for us, most strains are benign).

The two proteins look similar, but they have different functions. One study found that at least seven coordinated mutations must take place before one of these proteins can perform the function of the other, even a little bit.[24] Based on complex mathematical models, the estimated time for this is 10^{27} years. That's basically impossible; our universe is "only" 1.4 times 10^{10} years old. And this is bacteria, where we expect about 1,000 generations each year. There is no mathematical model for Darwinian evolution. The emperor of neo-Darwinian theory has no mathematical clothes.

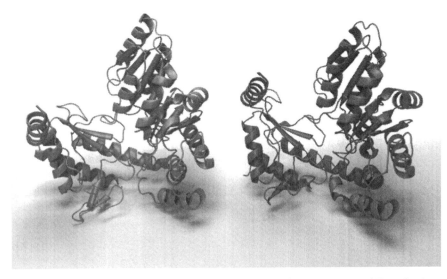

Illustration of proteins. Reproduced with permission of Ann Gauger and Douglas Axe.

To me, there are clear tensions between the concept that natural selection is some sort of inexorable force gradually improving organisms—some sort of constant, unrelenting life force contrary to the second law of thermodynamics (which I crudely summarize as the universe is heading toward maximum disorder), because of which organisms are constantly improving—and the sudden random beneficial mutations approach of punctuated equilibrium. Hundreds of scientists agree. Why should major changes in species always come quickly from small populations and not leave a fossil record? Where does the new genetic information come from, the precise coding to build new proteins and assemble them to form new biological systems and body parts? What makes all this superbly coordinated, mathematically improbable, new information pop up in a short period of time? Just as with the origin of life, the information problem is the Achilles' heel of neo-Darwinian evolutionary theory.

I promised to give you counterarguments, and here is one from Richard Dawkins and Stephen Jay Gould, among others. They say it is "false" that Darwin's theory rests on chance alone; they say it works by "cumulative selection." Say you start with a piece of gibberish and you want to transform it into meaningful text—you want to transform gibberish into "what fools these mortals be." You give the gibberish a spin—you randomly change letters, or mutate the DNA code in the world of life—and some of the resulting English letters/DNA letters may be correct. Not many, but perhaps one or maybe more. You hold on to any that are correct. You give the remaining English letters/DNA letters a second spin/mutation, and perhaps you get another one or more that are correct, and now you hold on to those also. You keep doing this. This process will converge in a reasonable period of time to yield the English phrase, or the DNA code, you are looking for. Dawkins and many others, with umpteen degrees and peer-reviewed articles to their credit, claim this "cumulative selection" drives Darwinian evolution.

To which I meekly respond, in the words of the great John McEnroe: "YOU CANNOT BE SERIOUS!" How does the gene know what it is looking for? How does it "cumulatively select" good letters and keep mutating only the bad letters? You get a designed product, and we're not supposed to notice the wizard behind the curtain running the show? You sneak intelligence in the back door but still claim the process is "natural"?

So to Dawkins and many others, "cumulative selection" works like this. You have an ordinary gene (let's call him "Joe") just sitting there, hiding out in the DNA, minding his own business. Not a lot happens in the genome. Sure there's replication, but the error rate on that is only one letter in a billion. But one day there is a mistake, and a DNA letter is changed by accident. Joe says, "Hey, that feels good, I might be able to use that someday" and decides to hang onto that change for future generations. It's a slow process, but Joe is patient. Sure enough, just 125,314 generations later, somehow, inexplicably, a string of DNA letters from Joe's buddy Hal gets inserted by accident when Joe is being replicated, and Joe says, "You know, that feels good too. I've got a hunch I might be able to use those extra letters someday, particular if another 27 changes are made in just the right spots in my DNA code."

Well what do you know, but it all works out perfect for Joe in just 72,437,197 more generations. Joe is able to hold on tight to all the changes that Joe likes and suspects he might use someday, and he is able to accept only those new changes that ultimately work just right with the old. Also, somehow, none of these changes affects the proteins that Joe is the code for in any negative way, so that they and Joe are fully functional through all of these generations right up to that special day, that day when the last "accidental" change is made and Joe transforms into a totally new gene with new capabilities. Joe is no longer ordinary Joe, he is now Malcolm, and Malcolm can build proteins Joe could never imagine. With Malcolm's proteins the organism has a major evolutionary advantage, and in the blink of an evolutionary eye Malcolm conquers the entire species.

Anyhow, that's neo-Darwinian theory. And, to quote another of my heroes, "celebrity" author Dave Barry (he does have a sewage lifting station in Grand Forks, North Dakota, named after him),[25] "I am not making this up."

I mean really. What a fairy tale! We are supposed to believe this? Genes don't "know" in advance which accidental mutations will ultimately prove useful. Genes aren't able to hold onto potentially "good" changes and prevent them from accidentally mutating also. If you switch the analogy from mutating DNA codes to mutating letters of the alphabet, you can use this theory of cumulative selection to transform text into any other text, even in a different language, by repeatedly copying the text and keeping only good mistakes. David Berlinski humorously suggests that all great literature, including "the Ulysses, mistakenly attributed to the Irishman James Joyce," was derived in this way by French monks, unable to read Spanish, copying "the Quixote."[26]

I have never seen such a patently absurd theory proposed by so many "educated" people. If it were any subject other than Darwinian dogma, where dissent is punished by career death, this nonsense would be torn to shreds. This is not belief. This is not reading the data in different ways. This is not a matter of interpretation. This is plain illogic. No scientist should be exempt from intellectual honesty. Any institution of "science" that cannot or will not reject this nonsense does not deserve our respect.

In the example above, while all the changes were being made, protein Joe was not affected until he transformed into Malcolm. I think almost anyone who has ever programmed a computer or worked on a complex piece of machinery knows that is a ridiculous assumption. According to University of Illinois biologist Tom Frazzetta, "The evolutionary problem is, in a real sense, the gradual improvement of an engine while it is still running."[27]

Junk DNA?

Let's go back to our analogy of mutations as changes to the letters and spaces in a paragraph. It's a fair analogy: proteins are built out of sequences of twenty different amino acids; text is built out of sequences of twenty-six letters and spaces and punctuation. A typical paragraph might have three hundred to four hundred letters/spaces/punctuation marks, a typical protein of a relatively simple organism might have three hundred to four hundred amino acids linked together. Just as a paragraph is a specified sequence of letters, spaces, and punctuation, a functional protein is a specified sequence of amino acids. The sequence also has to be folded correctly, and other information has to exist to produce that protein at the right time and in the right amounts, but we will ignore those bits of complexity here.

Neo-Darwinian theory claims that the DNA code for functional proteins mutates by chance into the DNA code for different functional proteins. Neo-Darwinian theory claims such mutations occur from combinations of the following as a result of accidental changes in DNA: (a) one amino acid is replaced by another, (b) one or more amino acids are deleted from the sequence, and (c) one or more amino acids (perhaps from another protein) are inserted. If we go back to the analogy of letters in a paragraph, these changes are equivalent to "mutating" a paragraph by combinations of (a) changing one letter/space into another, (b) deleting one or more letter/spaces, and (c) inserting text from another paragraph.

Neo-Darwinian theory claims that accidental, random, bump-in-the-night mutations in DNA coding change the amino acid sequence and continually create useful new proteins. So their theory, and believe me I am not making this up either, is that you sort of shuffle small pieces of the DNA, you mix it up in haphazard, unorganized ways, and out pops the information to build a new functional protein. And this happens over and over again, billions of times in the history of life, to produce all of the amazing species on Earth. And this does not happen just one at a time, but sometimes when you shuffle you get multiple new useful proteins that interact with each other just perfectly to create wholly new biological systems and body parts.

As we have seen, that theory is mathematically ridiculous. The monkeys aren't likely to slice and dice Cervantes into James Joyce, or Shakespeare into Tolstoy. But forget about the odds for now. If what they claim is true, then clearly most of the time during this slicing and dicing, during these accidental mutations of DNA, it doesn't work out very well. Most of the time, actually far more than 99.99 percent of the time, what you get is a piece of junk. A section of random, useless, "junk DNA."

If neo-Darwinian theory is true, then, as DNA is accidentally changed, we would expect to find, interspersed among various intact genes (intact sections of DNA that are the code to build a functional protein) and transformed genes (new coherent sections of DNA that build a new functional protein), a lot of random, useless coding. A huge amount of junk DNA.

That is exactly what die-hard Darwinists first claimed. When early experiments revealed that less than 2 percent of human DNA codes for proteins, the remainder was quickly but falsely labeled junk DNA. A 2003 article in *Scientific American* calls this assumption "too hasty" and "one of the biggest mistakes in molecular biology." The article added:

> The extent of this unseen genome is not yet clear, but at least two layers of information exist outside the traditionally recognized genes. One layer is woven throughout the vast "noncoding" sequences of DNA that interrupt and separate genes. Though *long ago written off as irrelevant because they yield no proteins,* many of these sections have been preserved mostly intact through millions of years of evolution. That suggests they do something

indispensable. And indeed a large number are transcribed into varieties of RNA that perform a much wider range of functions than biologists had imagined possible. Some scientists now suspect that much of what makes one person, and one species, different from the next are variations in the gems hidden within our "junk" DNA.[28]

A 2006 article in *Nature* magazine reported that scientists are using code-breaking methods to analyze DNA's layers of information:

> [R]esearchers now know that there are numerous other layers of biological information in DNA, interspersed between, or superimposed on, the passages written in the triplet code. Human DNA contains tissue-specific information that instructs brain or muscle cells to produce the suite of proteins that make them brain or muscle cells. Other signals in the sequence help decide at what points DNA should coil around its scaffolds or structural proteins. . . . [M]any stretches of DNA in humans and other organisms manage to multi task: a sequence can code for a protein and still manage to guide the position of a nucleosome.[29]

The junk DNA myth took a perhaps fatal hit in September 2012. Four hundred and fifty scientists worldwide, working on the Encyclopedia of DNA Elements (ENCODE) project to map the human genome, simultaneously released thirty major papers. The lead paper in *Nature* noted that they were able to "assign biochemical functions for 80% of the genome, in particular outside of the well-studied protein-coding regions."[30] "It's likely that 80 percent will go to 100 percent," stated one of the project's lead researchers. "We don't really have any large chunks of redundant DNA. This metaphor of junk isn't that useful."[31] The front page of the *New York Times* announced: "The human genome is packed with at least four million gene switches that reside in bits of DNA that once were dismissed as 'junk' but that turn out to play critical roles in controlling how cells, organs and other tissues behave."[32]

Now back to 1998, when the myth of junk DNA ruled. Some say that intelligent design doesn't make predictions, but it does. It predicts that scientists will continue to find design in biological systems. In 1998, Bill Dembski predicted the demise of junk DNA:

> Consider the term "junk DNA." Implicit in this term is the view that because the genome of an organism has been cobbled together through a long, undirected evolutionary process, the genome is a patchwork of which only limited portions are essential to the organism. Thus on an evolutionary view we expect a lot of useless DNA. If, on the other hand, organisms are designed, we expect DNA, as much as possible, to exhibit function.[33]

ENCODE has validated design, and it is a huge problem for Darwinists. Some attempt to dance around it. As late as 2009, die-hard Darwinist Richard Dawkins wrote: "[I]t is a remarkable fact that the greater part (95 percent in the case of humans) of the genome might as well not be there, for all the difference it makes."[34] Three years later, just six days after the ENCODE results were released, in a public about-face worthy of a master politician, he stated that the ENCODE results were "exactly what a Darwinist would predict."*

Exactly what a Darwinist would predict? Why? If Dawkins now means to suggest that small amounts of extra DNA confer a major evolutionary disadvantage, he is clearly wrong. While we humans have 3.2 billion letters of DNA, the marbled lungfish has 132 billion, and a rare Japanese flower, *Paris japonica,* has 152 billion.[35] There is no scientific evidence that extra DNA is an evolutionary disadvantage.

To me, ENCODE marks an important turning point in the ultimate demise of neo-Darwinism, although it may take years or even decades for our culture to appreciate its significance. You don't have to be a genius to see that, if all or almost all of our DNA serves a purpose, the theory that we arose simply from accidental mutations is utter nonsense. For this reason, many Darwinists ignore or deny the ENCODE results. "If ENCODE is right, then Evolution is wrong,"† admits one, using "Evolution" here to refer to neo-Darwinian theory. But ENCODE is not some fringe group. It is an international collaboration of 450 of the world's most respected scientists, working together, with no religious agenda whatsoever, "to build a comprehensive parts list of functional elements in the human genome."[36]

The ENCODE results make common sense. Anyone who has ever assembled a difficult item and spread parts across a garage or family room floor knows that it takes much more than the parts themselves to do the job. You need a lot of information on how and in what order to put the parts together. Now consider something almost unimaginably complex,

---

* This occurred in a debate with Britain's chief rabbi, available at http://www.youtube.com/watch?v=roFdPHdhgKQ.

† This quote comes from a presentation in Chicago in July 2013 by Dan Graur of the University of Houston. "Evolutionary Biologist." See http://twileshare.com/askq (accessed July 17, 2013). Graur describes himself on Twitter as "An Angry Evolutionary Biologist." See https://twitter.com/DanGraur (accessed July 17, 2013).

PUZZLES OF MACROEVOLUTION

such as our human bodies, with 30 trillion specialized yet interconnected cells. It takes a huge amount of information to put it all together and turn the systems in our body on and off when necessary.

Let's talk liver. Your liver is a three to four pound miracle. It processes food and cleanses the body of waste products according to daily cycles. In January 2013, scientists discovered that more than 3,000 "epigenetic switches" control liver functions.[37] An "epigenetic switch" is a section of DNA outside your genes (remember a gene is a section of DNA that contains the instructions to build one or more proteins). Epigenetic switches control "gene expression"; they tell your cells when to build proteins and how many to build. These 3,000 epigenetic switches are exactly what, until recently, was considered junk DNA. They are part of the "at least 4 million" gene switches discovered by ENCODE.

I think the efficient coding of DNA in multiple layers of information reveals design. Something is going on, something grander than haphazard mutations and natural selection. That something is scientific evidence of the existence of God.

We saw in chapter 10 that, out of all possible sequences of 150 linked amino acids, only one in 10^{77} can be expected to form a protein that can perform a specified function. Functional proteins are extremely rare in the space of amino acid sequences, and you cannot seriously expect that you can repeatedly "hop" from a protein that performs one function to a protein that performs a different function by accident. In the entire history of life on Earth, there have "only" been about 10^{40} organisms, and almost all were one-celled creatures. In the language of probability, you just don't have enough opportunities to expect to form by accident a single new functional protein, much less millions of new proteins. Many of the animals of the Cambrian explosion needed the complex protein lysyl oxidase to support their bodies; that protein is made up of "over 400 precisely sequenced (nonrepeating) amino acids."[38]

There has been zero serious scientific rebuttal to Doug Axe's one in 10^{77} estimate in 2004. An earlier 1990 paper by two MIT biologists estimated the odds of getting a sequence of 92 amino acids to perform a particular function as one in 10^{63}.[39] That's fantastically small. As we saw in chapter 8, a ball of marbles, one-half inch in diameter, extending out in all directions 50 light-years from Earth, 600 trillion (600,000,000,000,000) miles in diameter, has about 10^{60} marbles.

If you start with a protein that has a particular function, to some extent you can generally change some of the amino acids in its sequence and the protein will still perform that same function. But you can't rationally expect to "accidentally" mutate from one functional protein to another amino acid sequence with a specified new function. The odds of that are 1 in 10^{63} for 92 amino acid sequences and 1 in 10^{77} for 150 amino acid sequences, and the odds are far worse for longer amino acid sequences. Cumulative selection is a fairy tale.

Irreducible Complexity

As I read the literature challenging neo-Darwinian theory, I came across the concept of "irreducible complexity." The basic idea is that some systems are put together so that if you take away any of the parts, the system won't work. The classic example is a mousetrap. If you take away any of the platform, the spring, the hammer, the holding bar, or the catch, the rest of the pieces are useless. So the mousetrap is said to be irreducibly complex.

As I mentioned in the introduction to this chapter, Charles Darwin recognized that an irreducibly complex biological system could not be produced by natural selection:

> If it could be demonstrated that any complex organ existed which could not possibly have been formed by numerous, successive, slight modifications, my theory would absolutely break down.

And so we ask, are there biological systems that are irreducibly complex?

Advocates of design believe many biological systems have this quality. Here are two:

1. The bacterial flagellum—a molecular propeller that bacteria such as *E. coli* use to move. The bacteria flagellum resembles a rotary engine with up to 10,000 RPM. It has a variety of parts. The information to build these parts is typically coded by thirty to forty genes (some of these build the flagellum). It has been proven that if you take away any of these genes, the flagellum doesn't work. In January 2013, scientists discovered a species of bacteria with spectacularly

complex flagella—seven coordinated motors in a hexagonal array separated by gear wheels.[40] Bill Dembski told me the flagellum is the "poster child of intelligent design."

2. Blood clotting—blood clots inside the body can be deadly; uncontrolled bleeding is also fatal. When the skin is cut, a cascade of events, involving over two dozen factors, takes place to seal the wound. Blood clotting works pretty much the same for all mammals; it is highly conserved throughout biology.*

To me, these systems look irreducibly complex. Many scientists disagree, and suggest ways some parts of these systems could have formed by accident. The neo-Darwinian response is that all of the necessary parts arose in different systems where they played a useful role. It's a genuine debate, and part of the problem is that we don't know all the logical pathways to creating and assembling biological parts.

But even if some or all of the parts could have arisen elsewhere, you still need a tremendous amount of information to put the parts together. A mousetrap doesn't work unless the parts are put together just right. To me, these systems appear to be irreducibly complex. To me, that is scientific evidence of design.

There are many, many amazing biological systems. Some are noted in *Billions of Missing Links*,[41] a pro-design book by Geoffrey Simmons. One of his favorite systems is the attack spray of the African bombardier beetle. It carries separate storage tanks of hydrogen peroxide and hydroquinone: each is a relatively harmless chemical, but when mixed and a special "catalyst" added, they create a toxic stream. The bombardier beetle has both liquids plus the catalyst plus machinery to combine, aim, and fire this toxic stream at will. It can fire five hundred bursts per second, each burst with a velocity of sixty-five feet per second.

Recent evidence of design, and the wonder of life, comes from the loggerhead turtle. Adult loggerhead turtles are typically about a yard in length, and they are found in most temperate parts of the world. Some Pacific loggerhead turtles make the longest migration of any aquatic creature; their eggs are hatched in Japan, they "traverse the entire Pacific

---

* Michael Behe includes a chapter on this irreducibly complex process in his book *Darwin's Black Box* (New York: Free Press, 1996), pp. 74–97.

Ocean,"[42] six thousand miles to California, and then return to Japan, to lay their eggs on the same beach where they were hatched. How can they navigate so precisely across the Pacific Ocean and back?

In 2012, scientists announced that loggerhead turtles navigate by Earth's magnetic field.[43] They detect and use both the direction and the intensity of Earth's magnetic field to complete spectacular migrations, twelve thousand miles for Pacific loggerhead turtles, eight thousand miles for Atlantic loggerhead turtles, to the beach where they were born. When I commented in an online scientific discussion group that this was "amazing," I got this response from physicist Rob Sheldon:

> It's more amazing than perhaps you realize.
>
> The Earth's magnetic field is approximately an "offset dipole," or a bar magnet that is placed near the center of the Earth, but a tad closer to China. The direction of a dipole field tells you latitude, because the field is vertical at the poles, and horizontal at the equator. But the longitude, as this article explains, is harder to determine. The fact that it's closer to China causes the field over Brazil, or the so-called South Atlantic Anomaly, to be a bit weaker. So if one knows the direction of the field and its magnitude, one can fit it to the model to find one's latitude and longitude. Measuring the magnitude of the magnetic field is tricky, and "magnetometers" have been employed only since the beginning of the 20th century (while satellite magnetometry is still an actively developing field).
>
> For one thing, the magnetic field is highly variable, with daily changes caused by ionosphere, and random changes caused by aurora. My sub-specialty of magnetospheric physics has perhaps 20–50 people who do nothing but generate magnetic "indices" that attempt to describe this magnetic field. Since it changes all the time, an international committee publishes the "International Geomagnetic Reference Field" which is updated every 5 years, but must extrapolate until the next 5 year model because the Earth's field is continuously changing (some think it is because a chunk of crystallized iron in the Earth's core is rotating).
>
> The upshot is that only in the last 3 years can your Android or iPhone measure the magnetic field with a magnetometer, compare it to the IGRF2010, find the "down" direction with the accelerometer, and figure out which way it is pointed so that, for example, the constellation app can tell you the name of the star you are looking at. And the loggerhead somehow is born with a magnetometer and an accelerometer, and IGRF2010 pre-loaded. Amazing.

The 2010 version of the *International Geomagnetic Reference Field,* which Rob Sheldon referred to above as IRGF2010, is sophisticated software, "a series of mathematical models of the Earth's main field and its annual rate of change." The turtle doesn't just have machinery to accurately measure the strength and direction of Earth's magnetic field; it somehow has "software" built into its brain so it can adjust to changes. (Another study concludes that sockeye salmon use Earth's magnetic field to navigate thousands of miles back to the stream where they were born.)[44]

Note that, while your smartphone has been able to use Earth's magnetic field only in the last few years, loggerhead turtles, and possibly many other living creatures, have had this ability for millions of years.

I have a gut sense that the ability of the loggerhead turtle to navigate according to Earth's magnetic field must be, at some levels, an irreducibly complex system. It's hard to see how it could ever develop in a gradual step-by-step, purely Darwinian process. For example, simply knowing the strength of a magnetic field would appear to have no evolutionary advantage, yet that appears to be essential for loggerhead navigation. And how did the incredible "software" needed to navigate by Earth's magnetic field arise by chance?

A discussion of amazing biological systems could go on and on. Did you know Caribbean reef squid communicate by the colors, patterns, and textures on their skin?* I suspect there are thousands, perhaps millions, of irreducibly complex systems in living creatures.

Orphan Genes

Soon after scientists began to read the code in DNA, they discovered that many of the genes in each species had no family.[45] They also discovered that many of these so-called "orphan genes" play a key role in making that species unique, such as creating toxins in jellyfish and preventing freezing in polar cod.

---

* "In less than 30 milliseconds (yes, milliseconds!!), and in response to a visual cue, the squid can purposefully change its color and pattern to communicate with another squid, court another squid, stun their prey, or to try and ward off or confuse a potential predator. Many of the color/pattern displays include certain arm postures, as well. The squid can manipulate its color/pattern display in order to make it appear to move across their body.

Even more astounding is the fact that the squid can communicate two different messages to two different squid at the same time!" See http://sciencereasonfaith.com/caribbean-reef- squid-a-conundrum-for-neo-darwinian-evolution/.

As might be expected, the original consensus was that further analysis would solve the puzzle and that the ancestors of each such orphan gene would be found. But that has not happened. "Orphan genes have since been found in every genome sequenced to date, from mosquito to man, roundworm to rat, and their members are still growing."[46] A 2009 paper found, "Comparative genome analyses indicate that every taxonomic group so far studied contains 10–20% of genes that lack recognizable homologs [similar counterparts] in other species."[47]

Orphan genes can comprise one-third or even more of a genome. The leafcutter ant has "9,361 proteins that are unique . . . , representing over half of its predicted genome."[48] Leafcutter ants are found in South and Central American and in parts of the Southwest United States; they can build underground nests almost a hundred feet in diameter containing millions of individuals in just a few years. Other than human beings, they create the largest and most complex societies of all animals. Another study of seven ant species found 28,581 genes that were unique only to ants and not found in other insects. On average, each ant species contained 1,715 unique genes—orphan genes.[49]

According to neo-Darwinian theory, all genes should be descended from earlier models, and orphan genes with no recognizable ancestors shouldn't exist. But they do. And, so far, they have been found in all creatures.

Clearly, orphan genes are misnamed. Their ancestors are not dead or lost; their ancestors never existed. They should be called "designer genes."

Human Beings

"What a piece of work is a Man!" says Prince Hamlet in Shakespeare's play. Molecular biology confirms the obvious; Hamlet was right.

Debate over the origin of human beings may be evolution's bloodiest battle. I don't want to get into the history or politics of this fight. I want to go straight to the evidence of wonder. Again, let's pose the existence of God as a scientific hypothesis, and test that hypothesis using the tools of modern science. The Bible says God created human beings; neo-Darwinian theory says human beings arose from a "natural," undirected process. What scientific evidence is there that human beings were designed, and thus that the God of the Bible exists?

Apes do look sort of like human beings. The fragmentary fossil record suggests that during the last 8 million or so years there have been various species that could be considered transitional, including *Australopithicus afarensus* ("Lucy") 3.2 million years ago and *Homo erectus* ("Turkana boy") 1.6 million years ago. But similarity in design does not prove the process was unguided. Mustang and Taurus cars have similar designs, "and you could argue that they developed from a common ancestor, 'Ford.' But the similarities between these cars are the result of common design,"[50] not common descent by an unguided process. So similarities between the great apes and human beings could be evidence of unguided, Darwinian descent, or they could be evidence of a common designer. We need modern science to tell us which.

Molecular biology does that. To begin with, and here there is no dispute from Atheists or within the scientific community, human beings are indeed a great piece of work. We have seen that the instructions to build us are contained in 3.2 billion subunits, or letters, of DNA, enough to fill about a thousand thick volumes with fine print. Our 25,000 or so genes take up about 1.5 percent of these letters; as we saw earlier, a gene contains the code to build a protein, a biological machine part, and our human body uses "alternative splicing" technology so that pieces of a gene can be spliced in various ways to create the code for different proteins. An "average" human gene contains 27,000 letters of DNA, which uses on average about 1,300 letters to build a protein with a mean length of 430 amino acids, and it requires on average about 10 "splices" to put those 1,300 letters together.[51]

You may have read that only 1 percent of our DNA is different from that of a chimpanzee. That is not true. That figure looked only at genes and the machine parts/proteins they code for, not at how the parts are put together and operated. Consider the following from an article in 2007 on the "myth of 1%":

> Using novel yardsticks and the flood of sequence data now available for several species, researchers have uncovered a wide range of genomic features that may help explain why we walk upright and have bigger brains—and why chimps remain resistant to AIDS and rarely miscarry. Researchers are finding that on top of the 1% distinction, chunks of missing DNA, extra genes, altered connections in gene networks, and the very structure of chromosomes confound any quantification of "humanness" versus "chimpness." "There isn't one single way to express the genetic distance between two complicated living organisms," Gagneux [a zoologist at UC San Diego] adds.[52]

Human beings have important features that chimpanzees and other great apes do not share. A key physical difference is our ability to walk upright. In the 2012 book *Science and Human Origins,* Ann Gauger notes that this requires coordinated anatomical changes:

> To walk and run effectively requires a new spine, a different shape and tilt to the pelvis, and legs that angle in from the hips, so we can keep our feet underneath us and avoid swaying from side to side as we move. We need knees, feet and toes designed for upright walking, and a skull that sits on top of the spine in a balanced position. (The dome of our skull is shifted rearward in order to accommodate our larger brain and yet remain balanced.) Our jaws and muscle attachments must be shifted, our face flattened, and the sinuses behind the face and the eye sockets located in different places, to permit a forward gaze and still be able to see where to put our feet.[53]

At least sixteen features, coordinated anatomical changes from the great apes, allow us to walk and run efficiently.[54] They are a key part of what makes our bodies human and require hundreds or perhaps thousands of precisely coordinated changes in our DNA.

It is said that humans and chimpanzees shared a common ancestor perhaps 6 to 9 million years ago. That's less than 1 million generations. One million generations is hugely inadequate for neo-Darwinian theory.[*]

Human beings have twenty distinct families of genes that chimpanzees and other mammals do not have.[55] Each family has multiple genes. Again, and forgive me if you are getting tired of me saying this, each of these genes codes for one or more proteins, molecular machine parts that make human beings distinct.

The key difference between us and the great apes is not physical. The key difference between the species—the defining difference—is the much greater sophistication and capability of the human brain. In

---

[*] According to scientists at Cornell, the time for a specific mutation that makes a new DNA binding site to first happen and then to become fixed in a primate population has been estimated at 6 million years. R. Durett and D. Schmidt, "Waiting for Regulatory Sequences to Appear," *Annals of Applied Probability* 17 (2007): 1–32 (see p. 19 in particular). If the new binding site requires two new bases, the time is estimated to be 216 million years. R. Durett and D. Schmidt, "Waiting for Two Mutations: With Applications to Regulatory Sequence Evolution and the Limits of Darwinian Evolution," *Genetics* 180 (2008): 1501–1509. A DNA binding site is a piece of DNA that is eight letters long.

2011 scientists identified 198 orphan genes in humans, chimpanzees, and orangutans that code for proteins used in the brain.[56] These are all "young," in an evolutionary sense, less than 25 million years old. And 54 of these are solely human. So it appears that human beings have at least 54 new brain genes, new DNA coding to build sophisticated nanotechnology we use to think and reason. Remember that each gene codes for at least one protein, and most human genes code for more. Since the human brain is superior, it appears that these 54 new brain genes—with coding for at least 54 extra protein machines—have a beneficial effect.

It is surely an oversimplification, but in some sense our understanding today of how our brains work resembles the limited knowledge 150 years ago of how cells work. We thought then cells were "homogeneous globules of plasm," a jellylike goo that somehow made life possible. Today the common perception is that our brains are just collections of neurons that somehow send electrical charges that allow us to think. But when you do think about it just a little, the underlying programming and sophistication must be breathtaking. Consider memory, especially photographic, or "eidetic," memory. When I was a young adult, I could stare at a page, turn away, and later "read" the text. Even apart from such feats, exactly how is it, at a molecular level, that you can remember passcodes, quote poetry, or know where you left your car keys? (OK, it doesn't always work perfectly.)

The human brain is not a simple system; it is perhaps the most complex system in all life. Modern science is only beginning to understand how it works, and many of its most spectacular features (memory access, logical reasoning, spatial perception, subconscious function, etc.) are not understood at all on a molecular level. When I was a graduate student in theoretical mathematics, I would sometimes end my day unable to solve a problem, such as proving a theorem, and wake up the next morning knowing exactly how to do it. I think most people have similar creative gifts, such as in art, music, business, or other fields. Where does creativity come from? How can reason and creativity possibly arise from a gradual neo-Darwinian process?

Today a computer can be expected to beat the best human chess player, but that is because the computer can analyze millions of potential positions quickly, not because the computer has developed a new strategy or theory. We really don't know how to program creativity. How can we think the human brain arose by a blind, chance-based process when, at this point in history, the entire human race is unable to create a machine with true creativity?

I think these wholly new human genes that code for nanotechnology—human brain proteins/machines—are quite sophisticated. It is also quite possible, perhaps essential, that some of these machines are interconnected—in the sense that some of these new proteins may need one or more of the other new proteins to function properly. Somehow, within just a few million years, and perhaps as recent as within the last 100,000 to 200,000 years, a blink of an eye in the history of life on this planet, some or all of these 54 new molecular machines were created.

I just hit you with a lot, so take a breath and marvel. At least sixteen anatomical changes to walk upright. Fifty-four new genes with new brain technology. Less than 1 million generations to evolve by chance, where each "replication" has an error rate of only one DNA letter in a billion.

Take another breath. These are facts of science; they have been discovered by observation, experimentation, and logic. They are not religious facts. They prove that Darwin's theory of natural selection acting on accidental mutations cannot possibly be a complete explanation of how human beings were created. Darwin's theory has no mathematical clothes. Number, not religious belief, proves it to be false.

Sure, species are connected. But it's common design, not common descent merely through chance-based, unguided processes. Imagine you are digging up skeletons of cars, and you come across some you attribute to the species "Ford," from the first specimen you call "model T" that inexplicably appears in the fossil record around 1908 (although one of your colleagues claims to have discovered a Ford model dating back to 1903, but you can't find one, so you don't believe him), to a variety of later and more advanced specimens, including your favorite, a "Focus Electric" that you carbon date back to 2013. To what do you attribute changes in the species? And to what would you attribute that 1970 AMC Gremlin fossil? (Maybe design, but not necessarily intelligent design.)

Darwinists would like you to believe that modern DNA analysis establishes clear connections between species. What they don't tell you is that the connections are very different, and the relationships are very different, when you look at different families of genes.* They also don't

---

* Meyer's book *Darwin's Doubt* contains a good summary of the contradictory results of attempts to classify species by molecular relationships; see chapters 5 and 6.

tell you that some puzzles of macroevolution defy Darwinian logic. For example, about fifty species of mammals have an appendix, but these species are so diverse that the appendix "must have evolved separately at least 32 times, and perhaps as many as 38 times."[57] So some blind, unguided, and ultimately chance-based process produced the same organ dozens of times in different species, or a common designer placed it there for a reason. Bats and dolphins have hundreds of similar genes for "echolocation," a form of sonar.[58]

As we've seen, evolution is a complex and often confusing subject. Sometimes I want to put the science aside and enjoy the wonder. For that, I recommend a beautiful new video—*Flight: The Genius of Birds.*[59]

How did living creatures learn to fly? As the video explains, to literally get off the ground, birds have hollow bones, an improved cardiovascular system, improved muscles, feathers, wing design, and more. Compare an eagle to the most advanced jet. The jet is a triumph of human achievement, yet to me the eagle is a much more advanced piece of engineering. The eagle has greater aerodynamic flexibility, can obtain its own food, and can do other things the jet cannot do, not the least of which is produce baby eagles. And all of this "eagle technology" starts with a single cell.

I began assuming neo-Darwinian theory must be true. I learned the facts of science contradict it, and the emperor of neo-Darwinian theory has no mathematical clothes. When the paradigm breaks, and the wonder sparkles, an awful lot of college biology professors are going to look pretty silly. To me, the puzzles of macroevolution are the fifth wonder of modern science, the fifth of seven in our count to God.

My journey next took me back to physics, for a closer look at planet Earth. You may have read our galaxy has billions of planets like Earth. Let's look under that headline. I think Earth is special.

CHAPTER 13

Our Special Earth

Was Earth designed for life?

A review of habitable zones—for animals as well as microbes, and in the galaxy and Universe as well as around our sun—leads to an inescapable conclusion: Earth is a rare place indeed.

PETER WARD and DONALD BROWNLEE, *Rare Earth*, p. 33

IN MARCH 2009 the United States launched the Kepler space observatory, named after Johannes Kepler, the seventeenth-century German astronomer noted briefly in chapter 5. The Kepler space observatory measures tiny variations in the light from over 145,000 stars in our section of the Milky Way. Planets are detected by the periodic dimming of light as they pass in front of their host stars. Before it malfunctioned in May 2013, Kepler had detected over 3,000 planet candidates.

Our mass media translates these findings into a message that our Earth is not special. A January 2013 NASA news release claimed that one in six stars may have an Earth-size planet, which led to headlines of "17 billion" Earth-size planets in our galaxy.[1] Possibly true, but definitely misleading. It takes a lot more than being "Earth size" for a planet to be capable of supporting life, and especially for a planet to be able to support life over billions of years. Beyond the headlines, within the scientific community, there is a growing realization that Earth is special.

This insight is relatively new, largely in the last twenty-five years. In this limited area at least, the paradigm is beginning to shift. Modern science is revealing the wonder of planet Earth.

How special is Earth? It may not be the most critical part of the great debate, but to me it's one of the most interesting. The wonder of our universe created just right for life, and the overwhelming evidence of design in the most primitive form of life, surely can stand on their own. They would not be tarnished if perfect Earthlike planets were as common as grains of sand. But the scientific evidence suggests our Earth may be pretty close to unique, at least in our galaxy. In *The Privileged Planet,* Guillermo Gonzalez and Jay Richards explain what is special about Earth, and do the math. They estimate the odds of finding another planet as welcoming as Earth in our entire galaxy may be less than one in a hundred.[2] To me, the discovery that our Earth is special is the sixth wonder of modern science.

Of course, God could have created other wonderful planets suitable for humans. Perhaps there are "stepping-stones" out there to help launch humanity into the galaxy. But if you just look at the evidence from science, and assume no divine intent, it appears unlikely we will find another planet equally capable of sustaining life over billions of years in the entire Milky Way Galaxy.

The numbers and calculations in this chapter are not as overwhelming as in the fine-tuning and biology chapters. Here there is no "one in a zillion zillion" stunning scientific fact; here it is more the gradual accumulation of modest improbabilities. The calculations themselves are less precise, in part because the "science" of how and why planets form, and of what is special about the Earth, is still young.

In a way, the debate begins with Nicolaus Copernicus. As we saw in chapter 5, Copernicus suggested that Earth and other planets in our solar system revolve around the Sun. Some attempt to generalize Copernicus's revolutionary insight into a so-called Copernican principle. "It is evident that in the post–Copernican era of human thinking, no well-informed and rational person can imagine that the Earth occupies a unique position in the universe."[3] Others go further and falsely impugn to Copernicus a pronouncement of universal mediocrity, that there is nothing special in our Earth, in us, in life, or even in existence.

It is not position in the universe that makes Earth special. The planets, the Sun, and other stars do not revolve around Earth, and Earth is not the center of anything in that sense. But neither, it appears, is any other spot in the universe. All parts of the universe are expanding away from each other, and, as counterintuitive as it may seem, there is no center to this expansion. Our three space dimensions are expanding everywhere, like the two-dimensional surface of a balloon expands when inflated. So the universe does not have a "center," and it makes no sense to say that Earth is or is not at the center.

What is special about Earth is its ability to sustain life. Life, as we know it, can exist only when the right raw materials are present, only when protected from radiation and bombardment, and only in a "Goldilocks" zone that is not too hot and not too cold. Life thrives only when conditions are stable for a long period. In each of these ways, modern science tells us Earth is special.

The Right Galaxy

In the beginning, seconds after the big bang, the universe was mostly hydrogen (75 percent by mass), with some helium (25 percent by mass) and tiny traces of lithium and beryllium. The heavier elements—oxygen, carbon, iron, and so on—were created later, either forged in stars as they burned or squeezed together by catastrophic stellar explosions. Each atom in our bodies of these heavier elements was forged in the interior of a star. We are built from the dust of dead stars.

Heavier elements have chemical properties that make life possible. Oxygen can give or take two electrons rather easily; it interacts, sometimes very quickly, with many other elements and molecules. Flames burn in oxygen, and metals corrode as they "oxidize." Carbon is a stable platform for connecting atoms. When pressed into alignment, carbon becomes diamond, the hardest (most rigid) atomic structure known today. Other elements have other special properties. It has been suggested that twenty-six elements (including oxygen, carbon, nitrogen, phosphorus, potassium, sodium, iron, and copper) are important, perhaps necessary, ingredients of life on Earth.[4] Compared to the universe as a whole, our Earth has a high proportion of these heavier elements.

Cosmologists, persons who study the history and structure of the universe, call all of these heavier elements—elements other than hydrogen and helium—"metals." Chemists don't think of oxygen, nitrogen, or sulfur as metals, but to a cosmologist they are, and we will use that definition in this chapter. They were created inside stars.

Many of the first stars in the early universe were massive. The life cycle of a star depends mostly on its size. Bigger stars burn hotter and die faster. Our Sun has been around for 5 billion years and has another 5 billion years to go; it will burn a total of 10 billion years. A star 30 times the mass of the Sun will burn about 10,000 times as fast and have a stable life of only a few tens of millions of years. When a massive star dies—a star with an initial mass more than 8 times the mass of the Sun—it explodes. The explosion is called a supernova.

As the massive stars of the early universe burned, and then exploded, they seeded the universe with metals. But the process has not been uniform. Many galaxies did not and do not contain interstellar gas dense enough to form a lot of massive stars. These galaxies are metal poor, and they probably don't have habitable planets among their billions of stars. Many globular clusters are metal poor and unlikely candidates for life, as are also most small, irregular, and elliptical galaxies.[5] One estimate is that 98 percent of the galaxies in our part of the universe have a smaller percentage of metal than the Milky Way.[6]

The Right Distance from the Galactic Center

The metallic content of stars generally varies with their distance from the center of a galaxy. The greater mass density of the center gives birth to more massive stars, and as they die and explode over millions and billions of years their metals are distributed to younger generations of stars. The outer sections of the Milky Way have a lower density of matter, and they have had fewer massive stars. The stars and planets there are metal poor. Scientists now believe planetary systems like ours are unlikely to be found in the outlying "rural" areas of our galaxy.

The center of our galaxy has the necessary metals for life, but it is a dangerous neighborhood. A black hole, equivalent in mass to 3 or 4 million Suns, an entity so heavy and dense that not even light can escape, is believed to rule the center. The accretion disk surrounding it generates lethal radiation that would sterilize any life. Stellar explosions are also deadly. It is now believed that a supernova within 30 light-years or a more powerful gamma ray burst within 3,000 to 6,000 light-years would severely affect life on Earth.[7]

Gamma ray bursts are the most energetic explosions in the universe. They can produce more energy in a few seconds than our Sun will produce in its entire lifetime. The energy comes out in beams, perhaps from the poles of a collapsing star. What causes gamma ray bursts is not fully understood, but it is clear a direct hit can be fatal to all life. The statistics to date suggest that a galaxy like our Milky Way should experience a gamma ray burst every 100,000 to a few million years.[8] No known extinction event in the fossil record can be tied to a supernova or a gamma ray burst, but traces of excess iron-60 in the ocean floor from 2.8 million years ago may have come from a supernova in a group of massive stars called the "Scorpius-Centaurus OB association" when it was about 100 light-years from Earth.[9]

Our Sun is in the part of our galaxy best suited for life. Astronomers call that part the "galactic habitable zone," or GHZ. Only about 10 percent of the stars in our galaxy are in the GHZ.[10] We are about 27,000 light-years from the galactic center; the GHZ probably extends a few thousand light-years in either direction from our position. The metallicity—metal content—drops off fairly rapidly. In our region, it goes down about 5 percent for each 1,000 light-years from the center of the galaxy.

A Safe Path

The Milky Way is a spiral galaxy, with a central disk and curved, or "spiral," arms of stars coming out from the center. In a spiral galaxy, stars orbit the center—our Sun takes about 225 million years to do this, and has made about twenty circuits thus far.

Stars in spiral arms circle the galaxy at about, but not exactly, the same speed. Over many millions of years the patterns change. Our Sun is currently located on the inner edge of the Orion arm.

In addition to stars, spiral arms contain concentrations of interstellar gas. The gas accumulates in a sort of density wave—it piles up like automobile traffic. Dense gas leads to star formation. Supernovae tend to occur in spiral arms, because the relatively short lifetimes of massive stars prevent them from moving too far from their birthplaces before they explode.

Our Sun's path around the galaxy is nearly circular. This helps avoid traffic. Our Sun is also currently located near the plane of our galaxy, in a

Two spiral galaxies, NGC 5426 and NGC 5427, colliding 90 million light-years away, as seen through the eight-meter Gemini-South Telescope in Chile. Courtesy of NASA.

relatively calm area far from the busy and dangerous regions of the Milky Way. It appears that we have been in this calm area for at least several tens of millions of years, and perhaps for hundreds of millions of years. This stability—this extended period of calm—appears to have been necessary for the survival of advanced life on Earth. To see why, we need to look at another serious danger to life.

On June 30, 1908, a meteorite only 100 meters in diameter exploded in the atmosphere above Siberia. The blast was about a thousand times more powerful than the atomic bomb dropped on Hiroshima, Japan; an estimated 80 million trees over 803 square miles were destroyed.[11] It remains the largest impact event in recorded history. About 65 million years ago, a larger meteorite hit Earth near what is now the Yucatan peninsula of Mexico. It was six to nine miles in diameter—big, but smaller than billions of objects in our solar system. It released over a billion times the energy of the atomic bombs dropped in World War II, and many think it annihilated the dinosaurs.

Over the last 500 million years, there have been other mass extinctions, some of which are believed to have been caused by collisions. Worst was the Permian-Triassic extinction event of 252 million years ago, also known as the "great dying." It saw the extinction of 96 percent of all marine species and 70 percent of terrestrial vertebrate species.

Even today, after 5 billion years of collisions and a gradual thinning of large objects as they crashed into Jupiter and other planets, the danger is great. The asteroid belt between Mars and Jupiter has 211 objects more than 60 miles (100 kilometers) in diameter.* It has at least 700,000 or more objects 1 kilometer or larger in diameter.[12] In June 2013 NASA announced that it discovered the 10,000th "near-Earth object," and "there are at least 10 times that many more to be found."[13] They define a near-Earth object as an object that can come within 28 million miles of Earth, and the known objects range up to 25 miles in diameter.

Beyond Neptune lies the Kupier belt, which is like the asteroid belt but twenty times as wide and twenty to two hundred times as massive.[14] Pluto, with a diameter of 2,300 kilometers, lies in the Kupier belt. Another dwarf planet three times further out—Eris—may be larger.

A thousand times further out from Pluto, at the edge of the solar system, there is generally believed to be an Oort Cloud of trillions of comets. The asteroid belt and the Kupier belt are mostly in the plane of the solar system, but the Oort Cloud surrounds the solar system like a sphere. It is believed to contain "many billions" of objects 12 miles or larger in diameter.[15] The average distance of Earth from the Sun—about 93 million miles—is called an "astronomical unit," or AU. The Oort Cloud is a spherical shell of lonely, frozen comets, 50,000 to 100,000 AU out, up to one-third of the distance to the nearest star.

The problem is that, over millions and billions of years, very slight changes can shift objects into different orbits and create the danger of collisions. When Earth was very young, billions of years ago, collisions

---

* Enter ">100" in the search box for the asteroid diameter in the following web page: http://ssd.jpl.nasa.gov/sbdb_query.cgi.

were more frequent. During a period now called the Late Heavy Bombardment, about 4.1 to 3.8 billion years ago, Earth was bombarded by objects, perhaps mostly from the Kupier belt and the Oort Cloud. Some believe these comets brought the water now in today's oceans. So it has been critical for life that our Sun is in a relatively stable part of the galaxy. Nearby stars could disrupt the solar system and result in a collision with an object large enough to vaporize all life on Earth.

Since no major, civilization-threatening object has struck Earth in all of recorded history, it is easy to think the danger is small. But it is always there. I remember in 1997, the year after I moved to Martha's Vineyard, taking my family and driving a few miles to the beach to see Comet Hale-Bopp over the ocean. It was a magnificent sight, although I realized later the trip was unnecessary; the comet was plainly visible standing in our driveway. If Hale-Bopp had hit Earth it might have released one hundred times the energy of the collision 65 million years ago that killed off the dinosaurs, and extinguished all plant and animal life.[16]

The Right Sun

Some say the Sun is an ordinary star. That's not true. Our Sun is special in several ways. First, it's about the right size. Our Sun is relatively massive; about 95 percent of all stars have less mass. About 75 percent of all stars are so-called red dwarfs. Red dwarf stars typically have larger variations in the energy they emit, which would pose problems for life. Because red dwarf stars are much fainter, a planet with water in liquid form would have to be much closer, and the star's stronger gravitational attraction would likely cause the planet to become "tidal locked"—with one side always facing the star, like the planet Mercury around our Sun, and our Moon around Earth. A tidal-locked planet would have one side in extreme heat and the other in extreme cold, and it would be a poor candidate for life.

Our Sun is big, but not too big. Stars with more mass than the Sun burn out quicker. A star with 30 times the mass of the Sun burns out in a few tens of millions of years, and a star with 1.5 solar masses in about 2 billion years.[17] (Our Sun is about halfway through its 10-billion-year period of relative stability.) It is unlikely that more than about 10 percent of all stars have the right mass for life. By being in this favored 10 percent, our Sun is special.

Second, our Sun is a solitary star, which permits planets to have stable orbits. Massive stars like the Sun typically form in groups of two or three, and it appears that about two-thirds of the stars like our Sun, in our part of the galaxy at least, are part of binary or multiple star systems.[18] Stars in binary or multiple star systems may have exactly the right mass, but their planets would have chaotic orbits. The interaction of one or more other nearby stars would make stable, circular, Earthlike orbits less likely. There would also be more variability in the amount of energy received.

Third, our Sun is not part of a globular cluster, a dense collection of thousands of stars packed into a relatively small space. For example, globular cluster M13 has 300,000 stars in a space 160 light-years across. The nearest star to our Sun, Proxima Centauri, is over 4 light-years away. Globular clusters are poor candidates for life because of variations in heat and gravitational interactions that would disturb stable orbits. In addition, most globular clusters are metal poor.

Fourth, the amount of light energy our Sun gives out is relatively stable. Its output varies by only one part in a thousand over an eleven-year sunspot cycle.[19] Scientists aren't sure how long this stability has existed or how long it will continue.* A stable source of energy helps create a stable environment for life.

The Right Solar System

Stars, and accompanying systems of planets, form out of a huge collapsing cloud of gas and dust. As the cloud collapses, small amounts of angular momentum play a larger role, like a figure skater who spins faster when she pulls in her arms, and the cloud begins to spin faster and forms a disk. The center bulge of the disk becomes the star, and planets form further out. In the beginning, the planets are all revolving around the center star in the same direction, and their orbits are roughly circular.

---

* It is possible that our Sun had a special beginning, and its unusual birth blessed it with properties favorable to life. Some scientific evidence suggests that a supernova 5 billion years ago may have triggered the birth of our Sun. Perhaps for this reason, our Sun has an unusually high percentage of metals compared with the other stars at the same distance from the galactic center. S. Pfalzner, "Early Evolution of the Birth Cluster of the Solar System," *Astronomy & Astrophysics* (2013): A82.

Over millions and billions of years, this initial order can get messed up. One way in which our solar system is at least a little special is because all of the planets are still revolving around the Sun in the same direction as the Sun rotates. In some other systems, astronomers have detected planets with orbits in the opposite rotation to their parent star, probably because another star came close and pulled them off course, leading ultimately to "wrong way" orbits. This apparently did not happen to our solar system; it appears that no other stars have ever passed close. Again, the average distance of Earth from the Sun—about 93 million miles—is called an "astronomical unit," or AU. It appears no stars have approached within 150 AU, which is about five times the distance of Neptune from the Sun.[20]

Our solar system is special because the orbits of all of the planets are fairly circular, and spaced apart to reduce interference.

Jupiter, and to a lesser extent the other outer planets, often protect Earth from wayward objects. We saw that in 1994, when Comet Shoemaker-Levy 9 crashed into Jupiter and was swallowed up. Jupiter has 318 times the mass of Earth and acts like the vacuum cleaner of the solar system. Jupiter has helped to "clear out" dangerous Earth orbit–crossing objects. Computer simulations suggest that collisions between Earth and Oort Cloud comets, like the collision that ended the reign of the dinosaurs, would be thousands of times more frequent if Jupiter did not exist.[21]

The outer planets—Jupiter, Saturn, Uranus, and Neptune—have developed stable orbits in harmony. Saturn, for example, takes exactly twice as long to orbit the Sun as does Jupiter. According to computer simulations, this stability and harmony probably formed 700 million years after the birth of the solar system, when Jupiter moved closer to the Sun, and Uranus and Neptune moved much further out.[22] Neptune likely disturbed a then larger Kuiper belt, sending objects hurtling into the inner solar system and causing the Late Heavy Bombardment beginning about 4.1 billion years ago.

It appears that stable, well-ordered solar systems are rare, perhaps very rare.[23] Based on computer simulations, "conditions must be just right" to create a solar system like ours.[24]

The Right Place in the Solar System

Just as there is a Goldilocks zone with the right amount of metals for stars in our galaxy, there is a Goldilocks zone—not too hot and not too cold—for planets around their stars. Earth is near the inner edge of the solar habitable zone, where liquid water can exist.[25] As the Sun has evolved, and grown hotter and brighter over billions of years, the solar habitable zone has changed. Our Earth's orbit falls within the "continuously habitable zone," or CHZ, where the temperature has always been just right for liquid water. The CHZ is narrow; it begins just 1 percent closer in than Earth's orbit and extends perhaps 10–15 percent farther out. Outside that zone, without liquid water, complex life as we know it could not exist.

Earth's orbit is special. It is only 1.7 percent from being a perfect circle. From what we have learned to date, studying planets around other stars, this may be rare. Earth's near-circular orbit helps keep the amount of energy received from the Sun relatively constant.

The Right Moon

We now come to a feature of our planet that may be quite rare. It is the Moon. Let's look at why the Moon is unusual, and then why it is believed to have been absolutely critical to the development of life on Earth.

To many scientists, the Moon is something that shouldn't exist. There is still uncertainty as to how the Moon was created. The most popular theory, sometimes called the "big splat" theory, is that, perhaps 4.4 billion years ago, Earth collided with a very large object—about the size of Mars—and the collision resulted in a cloud of debris in orbit around Earth that ultimately collected to form the Moon. This theory may explain why the Moon has a small iron core and is much less dense than Earth; it proposes that the iron core of the object that hit Earth "kept going" in effect and merged with Earth's core. But some evidence, such as levels of radioactivity in Moon rocks, does not agree with this theory.* In addition, recent evidence suggests that both Earth and the Moon were born with water of identical composition already present, which seems to contradict both the collision theory for the origin of the Moon and the theory that the

---

* Recent studies show the Moon "has an identical titanium composition as the Earth." This would not be expected if the Moon was formed largely from debris from a different object. See *New York Times*, April 3, 2012, Science section, page D4. The current evidence shows that the Moon's "oxygen, chromium, potassium and silicon isotopes are indistinguishable from Earth's." "The Day the Earth Exploded," *New Scientist*, July 6–12, 2013, pp. 30–33.

water in Earth's oceans came from comets during the Late Heavy Bombardment.[26] So where did our Moon come from?

A new theory is that a natural nuclear reaction deep in the interior of Earth blew up the entire planet and that the pieces ultimately collected into our present Earth/Moon system.[27] This could explain the similarities in composition, but it is hard to see how the pieces came back together so nicely.

What is clear is that, compared to the size of its planet, our Moon is the largest moon in the solar system.[*] Our Moon is more than one-quarter of the diameter of Earth. To some scientists, the Earth/Moon system acts like a double planet.

The Moon keeps the axis of Earth's rotation at about the same angle to its orbit around the Sun. The Moon prevents Jupiter from causing large changes in the tilt of Earth's axis. Earth's equator tilts about 23.5° compared to Earth's orbit around the Sun. If this tilt were to change substantially, then the places on Earth that are now the North and South Poles would move toward the equator, with enormous climate change and disruption to any advanced form of life.

Also important to life are the tides, the changing force of gravity on Earth's surface as the Moon orbits. We can easily see tides in the ocean, but even "solid" ground has tide swings of about eight inches—the ground swells up as the Moon passes overhead. Through tides, the gravitational energy of the Moon stirs up the surface of Earth. Tides mix nutrients from the land into the ocean, to the advantage of marine life.

It appears that such a large moon is rare, perhaps very rare. If a collision or explosion 4.4 billion years ago created the Moon, there would have been a lot of dust. The Spritzer Space Telescope has found such dust in only one of about four hundred young stars.[28] Since very few collisions will result in a large moon, our Moon may be rare indeed.

In many ways, our Moon is almost too good to be true. It appears in the sky to be almost exactly the same size as the Sun. The Sun is about 400 times farther away and is about 400 times larger in diameter. This "coincidence" occasionally gives rise to total eclipses, when, for a few minutes, the Moon comes precisely between Earth and the Sun, as viewed from the right spot on Earth. Total eclipses have helped scientists understand our universe; experiments during total eclipses have measured the Sun's corona

---

[*] If Pluto were still a planet, its moon Charon would be the largest compared to its planet.

to learn about stars, measured the slowing of the rotation of Earth, and confirmed Einstein's theory of general relativity (by measuring how the light from stars near the edge was bent by the Sun's gravity). In *The Privileged Planet,* Guillermo Gonzalez and Jay Richards estimate the odds of a suitable large moon as probably not more than 1 in 100,000.[29]

The Right Ingredients

We are only at the beginning of understanding how planets form and what they are composed of, but Earth may have just about the right amount of various ingredients to sustain life. Let's start with water. Water makes up only 0.02 percent of Earth's mass, but oceans cover 71 percent of Earth's surface, to an average depth of about 12,000 feet. Twice as much water, just 0.04 percent by mass, would turn Earth into a water world, with little land for human life. Meteorites from the outer asteroid belt are about 20 percent water;[30] that's a percentage of water one thousand times greater than Earth's. So it seems to me that Earth's 0.02 percent could be set just about perfect for human life.

Then there's carbon. Our bodies are 18 percent carbon, yet carbon makes up only a small portion of Earth's mass, perhaps 0.05 percent.[31] Carbon is important for many reasons, including, as perhaps is becoming clearer each year, regulating global temperature. Some meteorites from the asteroid belt appear to have almost 4 percent carbon, or almost one hundred times the percentage of carbon on Earth.[32]

Water and carbon appear to be two key ingredients of life:

> Despite our best efforts to step aside from terrestrial chauvinism and to seek out other solvents and structural chemistries for life, we are forced to conclude that water is the best of all possible solvents, and carbon compounds are apparently the best of all possible carriers of complex information.[33]

Earth also has a large core of molten iron, which generates a strong magnetic field that deflects the "solar wind" and protects Earth's atmosphere from erosion.

Finally, Earth has plate tectonics. Earth's crust is divided into huge "plates" that collide and grind into each other. Plate tectonics may be the "central requirement for life."[34] Because of plate tectonics, land exists, and Earth is not a smooth globe uniformly covered with water. Because of plate tectonics, Earth's crust is constantly churning, which regulates the greenhouse effect of carbon dioxide. Because of plate tectonics, the temperature of Earth's surface has permitted liquid water for billions of years, and life has evolved. Earth is the only planet in the solar system with plate tectonics.

When you add it all up, I think the evidence is strong that our Earth is rare, and special. Earth is in the right galaxy, is located the right distance from the center of the galaxy, travels a safe path through the galaxy, has the right Sun, is in the right solar system, is in the right place in the solar system, has the right Moon, has the right chemical ingredients, and has plate tectonics. To me, these discoveries are the sixth wonder of modern science, the sixth of seven in our count to God.

Next we'll look at what I consider to be the seventh wonder of modern science: the mathematical and nonmaterial structure of the universe. Then we'll review the arguments for and against the existence of God. Last I'll share how I "connect the dots" after these thirty years.

CHAPTER 14

A Foundation of Thought

What is the universe made of?

In the beginning was the Word.

JOHN 1:1

This BOOK USES the term "number" in a very broad sense. It is meant to include all of mathematics. Not just "numbers" in the narrow sense like 1, 3, 17, $\sqrt{3}$, and so on, but all mathematical equations and concepts, all of the patterns and beauty of mathematics.

One feature of modern physics, that is almost taken for granted but is truly astonishing when you step back and think about it, is that number—fantastically complicated mathematics—has taken center stage. Centuries ago, you could perhaps say that number was useful only occasionally and in unexpected ways. Take Kepler's first law that the orbits of the planets are ellipses with the Sun at one of the two foci. Cute, but obviously of limited use, and not very helpful in understanding why things are the way they are. Then in 1687 Newton gave us his three classical laws of motion and his theory of universal gravitation. Newton discovered the system of the world, how things appear in our everyday human scale of reality, and he was able to describe it using nothing more than high school algebra. After Newton, we knew that number plays a central role in understanding how the universe works.

In Newton's physics there is a special state of rest. Einstein taught us this is an illusion and that only relative motion can be measured (this is part of his theory of special relativity). Einstein also taught us that we can't measure or detect any difference between the force of gravity and merely accelerated motion (this is the "equivalence principle," which underlies his theory of general relativity)—the laws of physics are exactly the same in each. These are symmetries of our universe, and the equations of general relativity have been described as a thing of dazzling beauty.

Einstein connected, using concepts of number, our three spatial dimensions with the fourth dimension of time, to build a new concept of a four-dimensional space-time. Einstein connected mass and energy with his most famous equation, $E = mc^2$ (energy equals mass times the speed of light squared). We now know that matter transforms into energy, and energy to matter, by mathematical formula, and both bend time and space according to beautifully symmetric concepts of number. Einstein illuminated deep mathematical connections among energy, matter, space, and time. For example, energy exerts a gravitational force.

There is a great deal more that could be said about the mathematical nature of the universe. English physicist Roger Penrose (who noted the extreme fine-tuning of the big bang described in chapter 8) says quite a lot in his book *The Road to Reality—A Complete Guide to the Laws of the Universe*. Near the end, on page 1033, he writes:

> The most important single insight that has emerged from our journey, of more than two and one-half millennia, is that there is a deep unity between certain areas of mathematics and the workings of the physical world.[1]

This journey from human-scale reality to an appreciation of the mathematical underpinnings of the universe appears to be accelerating. Recent discoveries suggest we have more to learn. Scientists have discovered that all of what we directly experience is only about 5 percent of the total mass and energy in the universe. About 27 percent is so-called dark matter,[*]

---

[*] It's definitely off topic, but if you want to be cool about dark matter, you had better check out the dark matter rap. http://www.astronomy.ohio-state.edu/~dhw/Silliness/rap.lyrics.

which has a gravitational field and mass but otherwise seems to be undetectable, at least so far. Dark matter provides gravitational scaffolding for galaxies; we now know that it is essential for the formation of galaxies and thus life as we know it. The remaining 68 percent is so-called dark energy, a mysterious force causing all galaxies to accelerate away from each other. We don't know much about dark matter and dark energy, but they may be hidden features of the mathematical structure of space-time. At least that's my theory. I hope the Nobel Prize Committee gives me full credit. (And it would be nice if they would hurry up; I'm not getting any younger.)

If there is only number beneath, what then of belief in God? Concepts of number are only ideas; they have no meaning as anything but ideas. Whose ideas could they be? I believe they are ideas in the mind of God. To me, the foundation, the underpinnings, of all existence may ultimately be ideas in the mind of God, a thought of our creator. This is my belief; I don't claim that modern science compels such a conclusion, but to me it is a wondrous harmonization of modern science with the Abrahamic concept of a creator God. Number, and perhaps only number, is the foundation beneath our reality. Number is the art of ideas. Whose ideas? To me, the mathematical nature of the universe is part of the seventh wonder of modern science, part of the evidence toward a universe created by thought.

This may sound strange, so let me clarify. I am not suggesting that everything in the universe reduces to a mathematical equation or numerical concept. Love, hope, joy, wonder, and so much more are clearly not just numerical.

I also am not suggesting that, because the foundation of the universe is mathematical, the thoughts of God control everything. I do believe we are given free will (as we will see, some think quantum physics and free will are connected). I believe God has used thought to create the foundation of the universe—interstellar space, and all matter and energy and forces—but has also created on that foundation creatures such as human beings with free will.

At this stage, I doubt you're convinced. Sure, you say, number and math are everywhere in physics and the sciences, but how else can we measure, compute, and predict? Maybe we haven't found the smallest particle yet, but there are particles, there are electrons and protons and neutrons. How can the foundation be only number, only thought?

I understand these doubts. But let's look closely at those microscopic and subatomic particles. Enter their reality, the reality of the quantum.

Quantum Physics

"Quantum physics makes me so happy. It's like looking at the universe naked," says Sheldon on *The Big Bang Theory*, a popular television show on CBS about young scientists. It's a great quote from a funny show, although I have absolutely no comprehension of why my wife says I sometimes remind her of Sheldon. I do not own a single one of his T-shirts. (In technical terms, the subsets of our T-shirts do not intersect.)

But Sheldon has it right in a key way. In quantum physics, you get underneath normal concepts of matter, space, and time. Quantum physics deals with the properties and behavior of the universe at very small scales. It is one of the most powerful and successful theories in all of science.

A key feature is the uncertainty principle. You can never exactly measure both the position of a particle and its velocity. There is always some uncertainty as to position, velocity, or both. This is not just a limitation on our ability to measure; it's a deeper truth. The truth is that subatomic particles don't have exact positions or velocities.* Their true nature can only be described by complicated mathematical equations that give probabilities of where a particle may ultimately be detected and what its velocity may be. This is the essence of Schrödinger's wave equation, first proposed by Erwin Rudolf Josef Alexander Schrödinger in 1935, after extensive correspondence with Albert Einstein.

The uncertainty principle holds that the universe is not deterministic, which leaves room for free will. You cannot predict what will happen from what has happened before. Antoine Suarez, a Swiss quantum physicist, philosopher, and bioethicist, and director of the Center for Quantum Philosophy, goes further. He suggests that quantum physics follows from free will.[2]

In quantum physics, ordinary reality at our human scale disappears and yields to number. Heavy math: Schrödinger wave equations, Hamiltonian operators, Klein-Gordon equations, Hilbert spaces, Eigen functions, Lie algebras, and more. Time and space do not permit discussion of these concepts. Go ask your mother.

In his miracle year of 1905, Einstein proposed that light comes in packets, called "quanta." This was another shock to the scientific community—

---

* More precisely, subatomic particles don't have precise momenta, the product of mass and velocity.

another triumph of Einstein's ability to think outside the prevailing worldview. The paradigm for hundreds of years had been that light is a wave, and the wave theory of light was used to create optical instruments and explain other phenomena, like the rainbow. But we now know that light comes in particles, individual "photons."

In high school I read of the debate over the "real" nature of light. Is it "really" a particle or a wave? The strange truth is that light is both, and it is neither. If that doesn't seem right, it is because our human reality is not the reality of subatomic particles. In fact, it gets weirder, much weirder, and it isn't just photons that are weird; it's all matter. To see how weird quantum physics really is, let's look at three different types of experiments—one-slit/two-slit, entanglement, and the quantum Zeno effect.

One-Slit/Two-Slit

Suppose we send a beam of light through a single slit. The result is a simple line of intensity. It's like we're shooting bullets through the slit.[*] So light is a particle, right?

Now suppose we send the beam of light through two parallel slits, as in the following diagram.

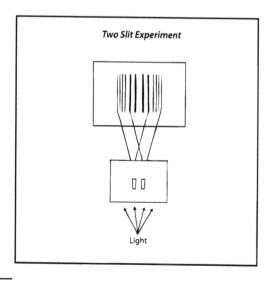

Two Slit Experiment

Light

---

[*] I assume here the slit is wide relative to the wavelength of the light.

If we were shooting bullets, we would get two parallel lines of intensity. But what we actually get is an interference pattern. The waves of the light beam interfere with each other, and where the wave peaks coincide is where the light is the brightest. It's exactly as if waves emanating from the slits were interfering with each other. So light is a wave, right?

Now we do the two-slit experiment again, but this time we send the photons through one at a time. What should we expect? If light is a particle, we should get two parallel lines, again like shooting individual bullets. If light is a wave, we're only sending one wave (photon) through at a time, so again we should get two parallel lines of intensity, right? Wrong! We still get an interference pattern. Somehow each individual photon "interferes" with itself. How can this be? As physicist Richard Feynman put it, "[T]he photon goes through one slit and it goes through both slits." Our human-scale reality is not the atomic scale reality of the photon.

This one-slit/two-slit experiment doesn't just work for photons of light. It also works with electrons and protons, and even with molecules. Buckyball molecules almost 500,000 times larger than a proton were found to exhibit wavelike interference.[3] All "ordinary" matter acts both like a wave and like a particle, and has this ability to "interfere" with itself.

Entanglement

In quantum physics two particles can be "entangled." This means they share the same quantum phase-state, which means they are EXACTLY alike. Here's one description:

> Quantum entanglement occurs when particles such as photons, electrons, molecules as large as buckyballs, and even small diamonds interact physically and then become separated; the type of interaction is such that each resulting member of a pair is properly described by the same quantum mechanical description (state), which is indefinite in terms of important factors such as position, momentum, spin, polarization, etc.[4]

When particles are so entangled, whatever you measure on one will be true of the other, and if you change one, you change the other. Before you observe or measure them, both particles will be in some unknown quantum state, with unknown qualities such as momentum, spin, and polarization. But when you observe/measure one, the other is affected, even if the particles are widely separated. In May 2012, scientists demonstrated quantum entanglement between photons over eighty-eight miles apart, between two Canary Islands.[5] Scientists believe there is no limit as to how far apart the particles can be. Theoretically, you could take one and move it to the other side of the universe, and if you did not observe/measure either, they would remain entangled. Einstein called this "spooky action at a distance."

That's plenty weird, but here's the really weird part. Observing/measuring one particle changes the other *instantaneously*. It happens *faster* than the speed of light; there is no delay whatsoever. This has been confirmed in experiments with measurements done before light could travel between the two particles.[6] I know this sounds like a mistake, but it's true. This faster-than-light action has been repeatedly proven. Entangled particles are connected outside of space and time. "There is no story in space and time that tells us how the correlations happen," states Nicolas Gisin, an experimental physicist at the University of Geneva in Switzerland. "There must exist some reality outside of space-time."[7]

Quantum Zeno Effect

One of the "paradoxes" attributed to the ancient Greek philosopher known as Zeno involves the motion of an arrow. The paradox is that, because at any instant when you look at the arrow it is at a particular place and not moving, the arrow cannot possibly move.

The quantum Zeno effect is similar, in that observing or measuring a system alters it. It's the quantum equivalent of a watched pot never boils. It has been experimentally confirmed, for example, that unstable particles will not decay, or will decay less rapidly, if they are observed. Somehow, observation changes the quantum system. We're talking pure observation, not interacting with the system in any way.

This same mystery occurs in the one-slit/two-slit experiment. Suppose we send particles one at a time through two slits, as before. We get an interference pattern; the individual particles somehow interfere with themselves. Now suppose we observe them going through the two slits. The interference pattern disappears; we get two parallel lines of intensity:

> The observer can decide whether or not to put detectors into the interfering path. That way, by deciding whether or not to determine the path

through the two-slit experiment, he/she can decide which property can become reality. If he/she chooses not to put the detectors there, then the interference pattern will become reality; if he/she does put the detectors there, then the beam path will become reality. Yet, most importantly, the observer has no influence on the specific element of the world that becomes reality. Specifically, if he/she chooses to determine the path, then he/she has no influence whatsoever over which of the two paths, the left one or the right one, nature will tell him/her is the one in which the particle is found. Likewise, if he/she chooses to observe the interference pattern, then he/she has no influence whatsoever over where in the observation plane he/she will observe a specific particle.[8]

Somehow, the act of observing changes the quantum system. Physicists speak of quantum waveforms "collapsing." What does that mean?

So we see that, deep down at the quantum level, our commonsense intuition as to how the universe works disappears. The "reality" of the quantum is not the reality we experience. What does this mean; what does it suggest about the substructure of the universe?

I think the answer is there; it's just too shocking for almost all scientists and philosophers to accept. It requires the greatest paradigm shift of all time, far greater than Earth revolving around the Sun, and far greater than the speed of light being constant. I suggest today's scientists are "educated" to ignore the obvious. But the answer is there, and the experimental facts cannot be denied. Here's how I see it.

Matter, space, and time are a constructed illusion. There is no such thing as a particle in the ordinary sense; particles are concepts constructed on a concept of space. (How else can you explain the one-slit/two-slit experiment, where "the photon goes through one slit and it goes through both slits"?) Space and time are a constructed illusion; there are connections outside of space and time. (How else can you explain quantum entanglement?) Perhaps thought alone is the foundation of the universe. (How else can you explain the quantum Zeno effect?) "The universe is immaterial—mental and spiritual."

That last quote is from Richard Conn Henry of John Hopkins University. In a 2006 review of a book on quantum physics, Henry writes:

In his Gifford lectures, very shortly after the 1925 discovery of quantum mechanics, Arthur Stanley Eddington (who immediately quantum mechanics was discovered realized that this meant that the universe was purely mental, and that indeed there was no such thing as "physical") said "it is difficult for the matter-of-fact physicist to accept the view that the substratum of everything is of mental character." What an understatement! On this fundamental topic, physicists are mostly terrified wimps.[9]

Six paragraphs later he continues:

> I really do not understand how it can be, that so little attention is directed to what is acknowledged to be the deepest discovery ever in human intellectual history: one that has changed our understanding of our own nature *far more* than did the Copernican Revolution.[10]

Physics led Richard Conn Henry to believe in God.

To put it plainly, Atheism/Scientism/Naturalism maintains that we exist as accidental creations of a material world, with no greater reality, a what-you-see-is-what-you-get reality. Quantum physics contradicts this view. Quantum physics suggests we exist, at least in part, in a what-you-think-is-what-you-get reality, and perhaps, just perhaps, in a what-God-thinks-is-what-you-get reality.

When you strip away preconceptions of matter, space, and time, quantum physics can be liberating. To quote Henry one last time: "It is more than 80 years since the discovery of quantum mechanics gave us the most fundamental insight ever into our nature; the overturning of the Copernican Revolution, and the restoration of us human beings to centrality in the Universe."[11] I suggest our minds are our connections to the ultimate—to God.

I know this may sound radical; I hope you can be open to the possibility. As they say, the facts are the facts. Number appears to be the foundation of the universe—the arrow of all of physics points to mathematics as the underpinnings of it all. Quantum physics defies conventional concepts of matter, space, and time, and replaces them with new realities of existence described by heavy math. If ideas, expressed as mathematical concepts, are the foundation of the universe, perhaps there is a great thinker. We have not proved the existence of the God of the Bible, but, again, we have taken a major step in that direction.

A Foundation of Thought

The equations of general relativity connect space, time, energy, and matter. The equations of quantum physics generally describe the microscopic level and below. Scientists have struggled for decades to reconcile the two theories, to come up with one unified theory, which some would describe as a theory of "quantum gravity." Thus far, they have failed. What has been learned is that if both theories are true, then space-time is discrete. If both theories are true, then there is a smallest unit of space-time.

Of course, we don't know if our current theories of general relativity and quantum physics are correct. We don't know how to extend general relativity to the level of the quantum, and it could well be, for example, that general relativity is only an approximation or a special case of a deeper theory, in the same way that the classical mechanics of Isaac Newton are an approximation of general relativity, a special case that works very well for velocities much slower than the speed of light. But both general relativity and quantum physics have withstood decades of testing.

But what if space-time is discrete? Space-time would have a granularity, and perhaps a fundamental unit. And, perhaps, that unit would be a mathematical concept.

My suggestion that the foundation of the universe may be made up solely of mathematical concepts is consistent with a branch of theoretical physics called "string theory." String theory proposes that everything in our reality—all matter, energy, space, and time—comes from fantastically complicated mathematical structures. Some propose that these structures have nine or ten dimensions, and that vibrations in the "strings" or surfaces of these structures create all of the effects of our reality. We don't experience nine or ten spatial dimensions, they say, because all but three are tightly rolled up at a subatomic level.

To me, quantum physics is the bridge between our reality and concepts of number. Hundreds of theoretical physicists are struggling to understand what those concepts could be.

Why should it be that everything in the universe obeys concepts of number—is described by elegant mathematics? How can quantum physics shatter ordinary concepts of matter, prove connections outside of space and time, and suggest that thoughts are supreme? To me, the combination of these two insights—toward an immaterial universe of thought—is the seventh wonder of modern science, the final step in our count to God. We do not live in a purely material world; we live in a world, a universe, of thought.

I believe God used thought, concepts of number, to create a universe capable of supporting life and human beings. The book of John in the New Testament begins: "In the beginning was the Word."[12] I believe this is consistent with all Abrahamic faith. I believe we and all of existence are thoughts in the mind of God. Quantum physics requires "a non-material agency outside of space-time," states the Center for Quantum Philosophy in Switzerland. "The material world emerges from non-material features."[13]

To illustrate this, I offer the following diagram. God uses number to create the universe, designed so that human beings can exist, creatures capable of both understanding number and detecting evidence of God. I call it the circle of existence.

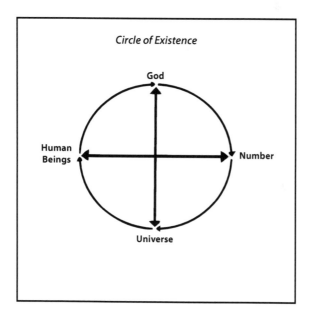

PART 3

Conclusions

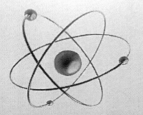

CHAPTER 15

The Logic of Belief

What's the argument here?

When you have eliminated the impossible, whatever remains, however improbable, must be the truth.

SHERLOCK HOLMES, in "The Sign of Four"

I HAVE GIVEN YOU seven wonders, seven pillars of scientific support for the existence of God. Let's look at the underlying logic. How exactly does science support belief? What are the basic arguments here, and what are the counterarguments?

Many, Atheists and even some theologians, will suggest I am arguing "from ignorance." They will suggest that most or even all of the "wonders" of this book, all of the incredible scientific evidence for the existence of God, are but "gaps" in our present knowledge. They will suggest that just because we don't currently know how something could have been created from nothing doesn't mean "God did it"; that just because we don't currently know how life formed doesn't mean "God did it," and so on. They will suggest I have fallen into a "God-of-the-gaps" fallacy.

It's a little insulting. It's like saying, "You're ignorant! Don't you know we will ultimately figure everything out with no need for God? Why can't you get with the modern world? It's a fallacy (you're ignorant) if you think that any observation, experiment, or reasoning could provide support for the existence of God. Don't you know God is a mere superstition?"

It surprises me how successful this God-of-the-gaps objection has often been in shutting down debate on design and the existence of God. I have come across respected scientists who, although they believe in God, somehow think this objection requires them to disown evidence of God. It's a tricky objection, because it assumes its own correctness. It assumes that science will ultimately provide a complete nontheistic explanation for all things. It assumes belief in "science-of-the-gaps." As I've said before, you can choose to believe in science-of-the-gaps or you can believe that the wonders described in this book are evidence of the existence of God. Your choice.

Some theologians have bought into this God-of-the-gaps objection, perhaps because they do not want to paint God into a corner, where every scientific advance reduces belief. Here's German theologian and martyr Dietrich Bonhoeffer:

> [H]ow wrong it is to use God as a stop-gap for the incompleteness of our knowledge. If in fact the frontiers of knowledge are being pushed further and further back (and that is bound to be the case), then God is being pushed back with them, and is therefore continually in retreat. We are to find God in what we know, not in what we don't know.[1]

I agree with Bonhoeffer that "we are to find God in what we know, not in what we don't know." That is exactly why I find the stunning results of modern science so compelling. We know that only intelligence produces information, and we have now found information in the universe and in life. If Bonhoeffer were alive today, I believe he would agree that we can know God through science.

This God-of the-gaps objection assumes that, as our scientific knowledge grows, the seven wonders of this book are shrinking. I strongly disagree; I think the contrary is true, molecular biology, for example, continues to reveal technology and design within life. A lot has happened since Bonhoeffer wrote those words in a Nazi prison in 1944. Each of the seven wonders of this book is based largely or totally on scientific facts and theories after World War II —the discoveries (1) that our universe was created, (2) of the fine-tuning of our universe, (3) of the impossibility of creating life by chance, (4) of DNA and the technology of life, (5) of the myriad puzzles of macroevolution, (6) that Earth is special, and (7) of quantum physics and the nonmaterial foundation of the universe. We may no longer see God in the rising of the Sun each morning, as many ancient civilizations did; now we see God in the creation of the universe and the resonance levels of carbon atoms. Instinctive awe over the wonder of life has been replaced by knowledge that all living creatures run off the same operating system and are built using a tremendous amount of information. Billions of monkeys typing for the entire life of the universe cannot rationally be expected to produce more than a short snippet of Shakespeare or functional DNA code.

I do not claim all gaps in our present scientific knowledge are evidence of the existence of God. I chose the seven wonders of this book carefully, based on what we know. For example:

Proposition 1: We know, from all of human history, and all of science, that only intelligence is capable of producing information.

Proposition 2: We know, from multiple scientific discoveries in the last few decades, that there is a tremendous amount of information hidden in the structure of the universe and in all living creatures.

Conclusion: The universe and life were designed.

This reasoning is focused and direct. It's a positive argument, based on finding in nature the type of information and complexity that, in all human experience, come only from intelligence. To me, design is the only plausible explanation for the creation of the functional nanotechnology in all cells, and the stupendous creation of human beings and the human brain. Yes, you can choose to believe in "cumulative selection," but I find that an illogical fairy tale that collapses upon even cursory examination.

We know so much more, from science—observation, experiment, and reasoning—that points to God:

- We know our universe was created, from Hubble's law about the movement of galaxies away from us, from the 1965 discovery of actual "relic photons" from the big bang, and from other evidence. A fundamental premise of science is that everything that comes into existence has a cause. What we know points directly to the existence of a supreme creator outside of space and time, to a first cause. It does not conclusively prove that God exists, but it certainly suggests that as a viable possibility, perhaps even the best conclusion, and one has to personally decide whether the alternative belief in an infinite multiverse (with its serious mathematical difficulties) that just exists for no reason is more plausible. The multiverse is pure science fiction; there is no scientific evidence that it exists, and there likely never will be any scientific evidence that it exists.

- We know our universe is fantastically fine-tuned for life. The laws of physics and dozens of constants of physics have been set unbelievably fantastically just right. This is information hidden in the structure of the universe, and it points to the existence of God. The alternative is to believe both in an infinite multiverse that just exists for no reason and that the laws and constants of physics can change. There is no scientific evidence that the laws and constants of physics can change.
- We know that all life is incredibly complex. At this time, there is no even "mildly plausible" theory for the origin of life by undirected natural means, and there is no expectation of any new law of chemistry or physics to explain the origin of life. There is information hidden in the structure of all life, design in fantastically complex assemblies of atoms and molecules, and it points to the existence of God. There is no Atheist explanation for the origin of life.
- We know the technology of life is billions of years old and more advanced in many ways than human programming and technology. This points to the existence of God. The Atheist belief that new functional nanotechnology arises from random combinations of atoms or even existing life has no mathematical clothes.
- We know there are unanswered puzzles in the emergence of wholly new species. To name just one, we now know that all creatures have designer (orphan) genes with no apparent relation to genes in other forms of life, and that those genes often help to make that creature unique. To be sure, Darwinists will fight to the death on this issue,

and speak rapturously of the power of natural selection, but "cumulative selection" is a fairy tale, and the fossil record and other facts of science do not agree with neo-Darwinian theory.

- We know Earth is special. Exactly how special remains to be determined, but right now it looks very special.
- We know from quantum physics that there are connections outside of space and time. We know that number, fantastically complex mathematical concepts and ideas, underlies all of physics and may be the foundation of existence. These known facts point to God.

As what we know increases, as the evidence grows, I think we have reasonably eliminated any pretense that materialistic views of reality are complete. What is left, as improbable as it may seem to some, is the existence of a transcending intelligence outside of space and time.

A second counterargument to the wonders of this book is that saying "God did it" has no explanatory or predictive value. Notice that this argument again assumes Scientism or at least a close cousin to Scientism. It assumes that only theories with explanatory or predictive value are worthwhile, that "science" in some narrow sense is everything, and that all focus should be on theories that can predict. One response to this is that intelligent design does predict we will continue to find evidence of God in the design of living creatures, and that prediction is being confirmed in laboratories around the world almost every day, such as the announcement in September 2012 by 450 scientists worldwide working on the ENCODE project that at least 80 percent of human DNA serves some function,[2] driving a stake through the heart of the myth of "junk DNA."

A more direct response is that I am not trying to predict. I seek meaning beyond prediction. I seek answers to the great questions. The great questions—such as why the universe exists and whether there is some type of greater reality and greater truth—are beyond science.

CHAPTER 16

Connecting the Dots

How does it all add up?

Wisdom has built her house, she has hewn her seven pillars. . . . Lay aside immaturity, and live, and walk in the way of insight.

PROVERBS 9:1–6

There are only two ways to live your life. One is as though nothing is a miracle. The other is as if everything is.

ALBERT EINSTEIN

WE HAVE JOURNEYED through seven wonders of modern science. In physics, we, meaning the collective wisdom of the entire human race over the last one hundred years, have learned our universe was created, our universe is fine-tuned for life, and our Earth is special. In biology, we have learned that even the simplest, most primitive life contains a staggering amount of information and that the origin of life requires the creation of hundreds of specialized proteins, including proteins to read and process DNA, and the simultaneous creation of the exact DNA code for the construction of those proteins. We have learned there is no evidence that the operating system of life, the central dogma of molecular biology, has evolved or changed in billions of years. We have learned the fossil record is devoid of the myriad transitional forms predicted by Charles Darwin and that the puzzles of macroevolution— the sudden emergence of wholly new species—cannot be explained by natural selection alone. We have learned that number, which I use to include fantastically complicated mathematical concepts, is not only our tool to understand the universe; it may be the foundation of the universe.

I do not mean to limit God to seven wonders. Some think it is a wonder the universe is capable of being understood, at least in part, by human beings. Others think it is a wonder the universe seems to be designed to allow us to learn about it. In many ways, the wonder of the universe seems unlimited; in the smile of a child, the colors of fish in the sea, the sound of a Mozart concerto. It's a wonder the places my car keys turn up.[*]

The seven wonders are a good fit with Abrahamic faith. When the book of Genesis came into final form, perhaps 2,600 years ago, it claimed the universe was created, the universe was made for life, Earth is special, and life was created. Each claim is supported by modern science.

Genesis begins with a good summary of how it all took place. The actual Hebrew word in Genesis 1 is *yom*, and one of its meanings is an indefinite period of time.[1] The Bible breaks down the creation of the universe and Earth and life into seven yoms to give ordinary people, including uneducated persons 2,600 years ago, a general sense of the awe of creation.[†]

I see broad, general consistency between modern science and the faiths of Abraham. There is room for wonder—awe, astonishment, surprise, and admiration—and for hope. To be sure, one can choose to believe that the universe is pointless and that everything in it, including Earth and all life and human beings, exists because of accidental, meaningless events. You can believe there are an infinite number of universes with different laws, features, and constants of physics. These are beliefs; they are not required by modern science. I think they are not consistent with the latest discoveries of science, with the proven facts of science, with the evidence of God. I find the paradigm of a pointless universe, so fervently embraced by the popular media and by the Atheist mind-set

---

[*] Fortunately for me, finding them is my wife's job. See Ell Marriage Contract Sec. 17(b)(3)(A)(iii) (Ninth Rev. Ed. 2011).

[†] To translate *yom* as merely a twenty-four-hour day clearly makes no sense, for what could a "day" be before Earth was created?

of many Western intellectuals, shockingly devoid of scientific grounding. To me, the facts favor Abrahamic faith, belief in a creator God.

I agree with John Mark Reynolds: "To use science to promote atheism is like using a man's child to prove he does not exist."[2] To me, the belief that we are here by accident is a superstition, and the claim that science supports it is the greatest fraud ever perpetuated on the human race.

We have put the great question about the existence of God to the test of science, and we have come to a place of wonder and mystery. Here's Robert Jastrow: "For the scientist who has lived by faith in the power of reason, the story ends like a bad dream. He has scaled the mountains of ignorance; he is about to conquer the highest peak; as he pulls himself over the final rock, he is greeted by a band of theologians who have been sitting there for centuries."[3]

What does science tell us of the creator? First, that there is a creator. By itself, this is "Deism." Deism is the belief that God created the universe for life but plays no further role. In the first cause, in the act of creation, and in the fine-tuning of the universe to make life possible, we have scientific evidence supporting a switch from Atheism to at least Deism. There is something outside our reality. To me, that is an inescapable fact of the big bang. You can call it God or you can imagine some blind universe-creating mechanism, but there is something beyond our reality, something outside of space and time. Deism is also supported by the fine-tuning of the universe. All agree that our universe is, or at least strongly appears to be, designed for life.

I think modern science takes us further, gives us strong evidence that our creator God has not abandoned us and is still present. About 4.4 billion years ago, our special Earth-Moon system was formed. About 3.5 billion years ago or perhaps even earlier, at or near the end of the Late Heavy Bombardment (largely a deluge of comets that may have left Earth with just the right amount of water), life was created. As we have seen, the simultaneous creation of the necessary machines of even the simplest form of life, together with the exact DNA code to build those same machines, is far beyond the reach of chance. Life may have formed in some primeval pond, but mere chance cannot be the explanation.

The puzzles of macroevolution—from the creation of dozens of radically new body structures and systems in the Cambrian explosion beginning 541 million years ago, to the creation of modern human beings, perhaps just over 100,000 years ago—are scientific evidence God is with us today. If you can somehow forget the battles over the evolution of human beings and look at it with fresh eyes, the evidence is overwhelming that all of the anatomical changes and intellectual capacities of human beings could not have arisen solely from blind mutations and natural selection. I believe natural selection was part of the process, but it was guided by God, perhaps at the quantum level. We were built by a master designer. Where else did those "designer genes" come from to make us human?

I see design in all creatures. How did the leaf-cutter ant get 9,361 unique proteins?

I'd like to conclude with some personal beliefs. This is what I believe in beyond science, beyond reproducible observation, experimentation, and logic; this is my innermost core of belief. And before I so share, I want to emphasize that I respect ALL religions, all persons on the Belief side of the great debate, all persons who believe in a greater reality. I believe there are many paths up the mountain.

I believe in a personal God, a God that is with us today. To me, the scientific evidence that an intelligence outside our reality designed the universe, designed life, and guided evolution through the eons up to the creation of modern human beings is compelling. I believe such a God would not abandon us.

I believe millions have personal experiences of God or of a greater reality. This may not be "reproducible" evidence in the scientific sense, but it is nevertheless impossible for me to ignore. I have met people who "knew" the instant a person they loved died, even though they were hundreds or thousands of miles away. I have met people who were visited by deceased loved ones in dreams. If you ask persons who work at hospices, I think they will tell you these experiences are common. Many keep these experiences to themselves; they fear others will think they are crazy. Clint Eastwood's 2010 movie, *Hereafter*, starring Matt Damon, explores how modern culture shuns such experiences.

I believe in life after death. I think we are all ideas in the mind of God, and if we are a "good" idea, God keeps us around, in some way known only to God. Some recent books and articles describe what it was like to be almost dead, brought to some heavenly place, and then brought back to life. I think many of these stories are true; I take great comfort in them.

Thousands of people have had near-death experiences, or NDEs. Skeptics try to write them off as delusions of stressed brain cells. One study of NDEs expected to find disjointed memories similar to our memories of dreams but found the opposite. "[N]ot only were the NDEs not similar to the memories of imagined events, but the phenomenological characteristics inherent to the memories of real events (e.g., memories of sensorial details) are even more numerous in the memories of NDEs than in the memories of real events."[4] In other words, memories of NDEs are more vivid and last longer than memories of "real" events. Orthopedic surgeon Mary Neal, whose NDE was caused by being underwater more than ten minutes during a kayaking accident in South America and who later studied other NDEs, says, "People who have been involved in a godly experience remember with clarity and constancy the details of the incident and vividly recall their emotions as though they had just occurred."[5]

One of the most comprehensive books on NDEs is *Evidence of the Afterlife: The Science of Near-Death Experiences,* by Jeffrey Long, MD, with Paul Perry.[6] Dr. Long obtained descriptions of over 1,300 NDEs. The results are astonishing—more than 95 percent of respondents stated that their NDE was "definitely real."[7] NDEs were reported by persons with no heartbeat and thus no brain activity.[8] NDEs were reported by persons under general anesthesia, even though "[i]t is medically inexplicable for anyone to have a heightened sense of consciousness while being at the brink of death."[9] The NDE descriptions from people around the world, from people in different cultures and different religions, and from children as well as adults, were strikingly consistent. Many reported out-of-body experiences, enhanced sight (including by blind persons!), meeting deceased relatives, and a sense of peace and love. Dr. Long states:

> By scientifically studying the more than 1,300 cases shared with NDERF, I believe that the nine lines of evidence presented in this book all converge on one central point: *There is life after death.*[10]

I recommend this book; the descriptions of NDEs will amaze you.

Not all reported near-death experiences involve heavenly beings; some survivors recall demons trying to drag them to hell.[11] Howard Storm was an "avowed Atheist" and self-admitted nasty person; in his book *Descent*

Into Death he describes demons tearing at and eating his flesh, and of being rescued by Jesus. He recovered, quit his job, and entered divinity school.[12]

I don't know what to believe about hell. Perhaps, if God doesn't think you are a "good" idea, then when you die, perhaps you cease to exist, and that separation from God is hell. But the Bible speaks of hell, Jesus spoke of it, and some of the near-death experiences describe it. Jesus used the word "Gehenna," a place outside Jerusalem where garbage was burned and criminals were buried. From this comes the medieval image of hell as a place of fire and brimstone.

Heaven or hell, the near-death experiences all suggest our consciousness continues in some form after death—we do have some type of eternal soul. Consciousness may possibly be like plugging into the Internet; our brains can partially tap into it, but there is a lot more than our brains involved, just as the Internet is a lot more than the computer in front of us. It may sound crazy, but it could explain near-death experiences. Here are the words of John Eccles, a Nobel Prize–winning neuroscientist who studied consciousness:

> I maintain that the human mystery is incredibly demeaned by scientific reductionism, with its claim in promissory materialism to account eventually for all of the spiritual world in terms of patterns of neuronal activity. This belief must be classified as a superstition. . . . We have to recognize that we are spiritual beings with souls existing in a spiritual world as well as material beings with bodies and brains existing in a material world.

I believe the New Testament is essentially an eyewitness account of the life and ministry of Jesus Christ, and of the early church following his death. As others have pointed out, Jesus doesn't give you a lot of wiggle room—you either have to conclude he was crazy or that he was actually God as he claimed. Excellent scholarship supports the veracity of the New Testament; an easy read is Lee Strobel's *The Case for Christ: A Journalist's Personal Investigation of the Evidence for Jesus.* Of all the facts surrounding the life and death of Jesus Christ, what I find most convincing happened in that upper room following his death. The disciples were in an upstairs room; they locked the doors in utter despair and abject fear, their leader had just been crucified, and they were sure they would be next. Then something happened, something so powerful that they became missionaries and created the Christian Church. That is all the proof I need of the truth of the resurrection. Jesus appeared in that upper room after his death.[*]

I believe God embraces all humanity, not just Christians. Some believe the New Testament damns those who do not believe in Jesus Christ—such as where Jesus says, "No one comes to the Father except through me."[13] That is not how I interpret the Bible. Just three verses later Jesus states: "Anyone who has seen me has seen the Father."[14] Jesus then asks, "Don't you believe that I am in the Father, and that the Father is in me?"[15] I think the point here is simple: if you get to Jesus, you get to God, and if you get to God, you get to Jesus. God and Jesus are the same; there is no "Jesus gateway" you have to pass through to get to God. I refuse to believe God would punish a person simply because he or she was not raised in the Christian tradition or would punish those who lived before Jesus was born or had no ability to learn of Jesus' existence. I find such suggestions to be un-Christian. To me, Jesus was clear that God loves us all.

I go to an Episcopal church. I was raised a Congregationalist. After my son was born, when my wife dragged my then Atheist behind back to church, we agreed we would visit and compare all the churches in our area. It just happened that the first church we went to was Episcopal, and, as I described in chapter 3, we liked it so much we skipped the others. The Episcopal Church is a good fit for me; it is rich in tradition and liturgy. I appreciate the intellectual freedom it affords, and I am proud of its inclusion of all members of our society. But I certainly don't think Episcopalians are necessarily better, or even better on average, than members of other Abrahamic faiths. The people I find most devout may be the Muslim cab drivers of Washington, DC. Again, I believe there are many paths up the mountain.

I do not claim religious believers are necessarily more moral than, or otherwise superior to, Atheists and Agnostics. One is reminded of words attributed to St. Augustine; the church is a hospital for sinners, not a museum to saints.

---

[*] Another book on this subject is N. T. Wright, *The Resurrection of the Son of God*, Vol. 3 of *Christian Origins and the Question of God* (Minneapolis, MN: Fortress Press, 2003), but its 740 pages are definitely not an easy read.

I believe the misuse of religion historically and today is an abomination. People misuse religion to control other people. That is a special kind of evil, and it is one reason why many reject religion: they perceive religion as being fractious and mean. It is easier to get people to fight or even kill themselves for your cause if you can persuade them that "God" wants them to do it. This misuse of religion is rampant in history, from the Crusades to Northern Ireland to al-Qaeda, and many other modern examples. But it is not cause to give up on faith or religion. The greatest atrocities in history have taken place when religion is shut out—Nazi Germany, Stalin's purges, Mao's Great Leap Forward, the killing fields of the Khmer Rouge. These bloodbaths occurred when secular rulers took the place of God. Killing in the name of God is a terrible thing, but far more have been killed in the name of man.

Which brings us to why God would allow this to happen, and more generally why God would allow so many horrible, broken conditions to exist—from cancer to crime to addiction to starvation to war and on and on. As I mentioned in chapter 1, others have addressed this subject eloquently, but to me it basically comes down to a decision by God to give us free will.

In *Bruce Almighty*, a wildly successful 2003 movie comedy, Bruce (played by Jim Carrey) is a TV reporter who complains to God about his problems, and God lets Bruce play God, except Bruce can't tell people he is God and can't alter free will. Bruce tries to make everybody happy, but screws up and causes problems that lead to riots, and loses his girlfriend (played by Jennifer Aniston). Bruce ends up complaining to God about how impossible it is to get people to love you if they have free will. God (played by Morgan Freeman) responds: "Welcome to my world."

I suspect this conveys a deep truth, that an unavoidable result of free will is that we have the power to ignore God and to hurt each other in horrible ways. When I listen to people complain about why the world is the way it is, and why God did it that way, I think it all basically comes down to their belief that, if they were God, they would do a better job. I doubt that's possible. I think free will requires us to accept responsibility and not blame everything on God. God has given us a universe of wonder, and it is up to us to make it a wonderful place.

And consistent with the concept of free will, and of giving us a choice, I think the existence of God is supposed to be a riddle. I think we are supposed to struggle with the puzzle. If the existence of God were too obvious, then we really wouldn't have a choice. It is a grand riddle, the "great riddle," and today, with modern science, we are better equipped than ever to understand it. I think we are supposed to use the gift of human reasoning to detect the existence of God. I think we are designed to do that.

Perhaps we are not capable of fully understanding why the universe is the way it is, and in particular why suffering exists. Is it a human conceit to think that we can grasp existence? Does the ant, or the shark, or the hummingbird understand its world? Of course not, yet they go about their business just as they are intended to do. Human beings are more, much more; we have been blessed with Godlike powers of reasoning, and we thirst for understanding. But perhaps there are truths beyond human understanding.

When people say they can't believe in a God who would allow so much suffering to exist, they are assuming suffering is pointless. I'm not sure of that; there have been times in my life where I have suffered, but I was better in the end for it. I have had horrific bouts of cluster headaches for forty years, which each time destroyed my ability to function yet each time left me with a greater appreciation of life. Senator John McCain suffered horrifically as a prisoner of war in Vietnam for over seven years; in solitary confinement for years at a time, repeatedly and brutally tortured, he led a life that most of us would find beyond endurance. Yet he has said he would not trade that experience for anything, because it was the basis for what he became. When the waters are rough, we do tend to think about the great questions, and we often draw closer to God. Often no real adversity means no real character, and that is why we tend to think of spoiled children, or overly indulged celebrities, as vacuous and superficial. The assumption, the belief, that all suffering is pointless, seems wrong.

What then of the suffering of hell? Some, like C. S. Lewis, suggest that hell is "the greatest monument" to the freedom given to us by God, to free will. Free will must include the ability to turn away from God and to deny the existence of God. In Lewis's words:

> There are only two kinds of people—those who say "Thy will be done" to
> God or those to whom God in the end says, "Thy will be done." All that

are in Hell choose it. Without that self-choice it wouldn't be Hell. No soul that seriously and constantly desires joy will ever miss it.[16]

I would refer you again to Timothy Keller's excellent book *The Reason for God: Belief in an Age of Skepticism,* for a thorough discussion of these difficult questions of faith.

I have taken pains in this book to be inclusive of the other Abrahamic faiths—Judaism and Islam. There are three reasons for that. First, I believe there are many paths up the mountain. Second, from the viewpoint of science, all of the wonders of this book point equally to all three Abrahamic faiths; you cannot say, for example, that the technology of DNA would lead you to believe in Christianity over Islam. But third, and most important in my view, is that many of us on the Belief side of the great debate tragically fail to cherish and embrace our brothers and sisters of other faiths.

We Believers need to wake up and see the world the way it is. The most significant battle of our generation, and for our children and our children's children, is not Islam versus Christianity; it is Scientism versus Belief. We Christians do not have the only pathway up the mountain to God. I believe we are very fortunate to have a path illuminated by the light of Jesus, a path where, once you step on it with the intent of going to the top, you are there, you are saved. (To me, Christianity is sort of like a fast escalator to God.) I believe Christianity is a true path, and I believe it offers reassuring truths that other paths do not share, but it is not the only path.

Just as many scientists are locked in a paradigm that shuts out God, many religious leaders are locked in a paradigm that shuts out other Abrahamic faiths. My slim encounters with Muslims, mostly brief conversations with cab drivers in Washington, DC, suggest to me that Muslims may on average be more devout than Christians. They are threatened, as we should be too, by the strong and increasing bias against Belief within our society. How did we in the United States ever get to the point where legitimate scientific concerns about Darwin's theories cannot be published in high school textbooks? How did we acquiesce in academic institutions turning their backs on founding religious principles? Why do we not challenge museums that put a single fossil on the wall and claim it validates Darwin's theories? How can fairy tales like cumulative selection be considered science?

If we are going to make the world a better place for our children and their children, then all Abrahamic faiths need to unite, and the sooner the better in my view, to at least stop poisoning young minds with Anti-Faith claims falsely made in the name of "science." We must use science as the great tool it is to help us understand the truth, and we must not be deceived by the hidden agenda of Scientism.

Let me make three points clear. First, I am not suggesting we ignore the fanatics of al-Qaeda. They are murderers hiding in religious clothes. I thank God for our military and all who protect us.

Second, I am not suggesting religious control of government or the media. All I am suggesting is free and open debate, and that we recognize Scientism as a belief. While we focus on the fanatics of al-Qaeda, the fanatics of Scientism have gained control. Can we think there are no consequences when children are falsely taught that science contradicts religion? Can we think there are no consequences with the popular paradigm of a pointless universe?

Third, I am not suggesting all paths up the mountain are the same. I am just suggesting that the biggest threat to each of the Abrahamic faiths is Scientism.

We use public money to hand high school students biology textbooks with Anti-Faith information we know is false. We have known for decades that the Miller-Urey experiment did not reproduce conditions on the early Earth and that the mere existence of amino acids does not solve the information problem of the origin of life. We have silently acquiesced to the fanatics of Scientism. We are ever vigilant against fanatics of religions we recognize, but, by hiding in the clothes of science, the fanatics of Scientism are invisible to us. They have gained control over much of our culture, and they seek to shut out the wonder. They falsely claim that evidence of design, in the universe and in life, is not science. It is science; it is just not Scientism.

When we deny our children the true facts of science, our children are vulnerable to claims that the universe is pointless and that life is but a chemical accident. We watch in horror as they abuse themselves with drugs and alcohol, and abuse each other. We watch them walk into public places with automatic weapons and start shooting. We ask "where is God" when it happens. I think God asks why we withhold and deny evidence of design. "For since the creation of the world God's invisible attributes—his eternal power and divine nature—have been understood and observed by what he made, so that people are without excuse."*

Believers are the majority of those living in the United States. We need to embrace legitimate science and legitimate scientific debate. We need to embrace our brothers and sisters in other Abrahamic faiths. There is danger in statements like "Christianity is the one true religion." It is divisive. If the question is whether the fundamental claims of Christianity are true, whether Jesus was the Son of God and rose from the dead, I say definitely yes. But if the question is whether the other Abrahamic faiths are false, I say definitely not. The best religion is the religion that gets you closest to God. I think Jesus would agree with that.

Thanks for sharing my journey through number, universe, and God. It's been an amazing trip. May your journey find wonder, and may your mind and your soul be open to the science of belief. I hope I've convinced you that science is a wonderful tool, and that we can use it to count to God.

---

* Romans 1:20 (International Standard Version).

SUGGESTED READING

There are hundreds of books that explore the themes of this book. Here are seven to get you started.

On the science of belief:

1. Lee Strobel, *The Case for a Creator: A Journalist Explores Scientific Evidence That Points to God* (Grand Rapids, MI: Zondervan, 2001).
2. Ann Gauger, Douglas Axe, and Casey Luskin, *Science and Human Origins* (Seattle: Discovery Institute Press, 2012).
3. Stephen C. Meyer, *Darwin's Doubt: The Explosive Origin of Animal Life and the Case for Intelligent Design* (New York: HarperOne, 2013).
4. William B. Dembski, *The Design Revolution: Answering the Toughest Questions about Intelligent Design* (Downers Grove, IL: InterVarsity Press, 2004).
5. Guillermo Gonzalez and Jay W. Richards, *The Privileged Planet: How Our Place in the Cosmos Is Designed for Discovery* (Washington, DC: Regnery Publishing, 2004).

On religion generally:

6. Timothy Keller, *The Reason for God: Belief in an Age of Skepticism* (New York: Riverhead Books, 2008).

On near-death experiences:

7. Jeffrey Long with Paul Perry, *Evidence of the Afterlife: The Science of Near-Death Experiences* (New York: HarperOne, 2010).

APPENDIX A

Exponents

THIS BOOK USES mathematical shorthand to write very large and very small numbers. This shorthand is called "exponents." For example, a one followed by a hundred zeros is written as 10^{100}. Written this way, the 100 means we multiply one hundred "10s" together. So 10^1 means multiply one 10 together, which is just 10. We have $10^2 = 10 \cdot 10 = 100$, $10^3 = 10 \cdot 10 \cdot 10 = 1,000$, and so on. Exponents "work" with other "bases," not just 10. We can write $2^1 = 2$, $2^2 = 2 \cdot 2 = 4$, $2^3 = 2 \cdot 2 \cdot 2 = 8$, and so on. For example, 3^4 means multiply four threes together ($= 3 \cdot 3 \cdot 3 \cdot 3$), which is 81.

Exponents may seem simple, but they are powerful. To begin with, they turn some multiplication problems into addition problems. If two numbers that have the same base are written with exponents, we can multiply them together just by adding the exponents. For example, we can multiply 100 ($=10^2$) times 1,000 ($=10^3$) just by adding $2 + 3 = 5$. $10^2 \cdot 10^3 = 10^5$. If you think about it a minute, this shouldn't be surprising. If you multiply two "tens" together, and then multiply that by three more "tens," you have five "tens" multiplied together. In this same way, $10^3 \cdot 10^7 = 10^{(3+7)} = 10^{10}$; $10^{14} \cdot 10^{17} = 10^{(14+17)} = 10^{31}$; $2^5 \cdot 2^{13} = 2^{(5+13)} = 2^{18}$, and so on.

A second reason why exponents are powerful is that we can expand the idea. Let the letter n stand for, be a "marker" in our mind for, any number not equal to zero. By writing $n^0 = 1$; $n^{-1} = 1/n$; $n^{-2} = 1/(n \cdot n) = 1/n^2$; and so on, we can expand our idea of exponents to include zero and negative integers as exponents. And our trick of turning multiplication into addition still works:

$$n^2 \cdot n^{-2} = (n \cdot n) \cdot 1/(n \cdot n) = 1 = n^0 = n^{(2+-2)}.$$

We can expand our ideas of exponents further, and use any number as an exponent. Exactly how this works and what it means are beyond the scope of this appendix, but it is a powerful tool. For any positive number y, there is a number x such that $y = 10^x$. If y is less than one, then x is negative. In this equation we say that x is the "logarithm" of y. Using logarithms, you can multiply numbers together just by adding their logarithms and then reversing (inverting) the process. For hundreds of years, long before there were computers or even calculators, this concept—this idea—made it possible for sailors to do the calculations needed to navigate around Earth, and made possible much of the industrial age.

APPENDIX B

Infinity

THIS APPENDIX is a mind-bending trip into the world of infinity. Infinity is a different kind of number. Stated another way, infinity is a different kind of idea. Infinity often doesn't behave like other numbers, and that's part of what makes it alien, exotic, and fun.

Suppose—imagine—there is a motel with an infinite number of rooms. Let's number them Room 1, Room 2, Room 3, and so on, so each room has its own natural number. Assume each room is identical and can hold exactly one, and only one, person. It is, let us also assume, the only motel in town.

Now suppose the Shriners want to have their annual convention at this motel. Suppose—remember the name of this appendix is "Infinity"—there are an infinite number of Shriners, and we can line them all up and give each a natural number so that we can tell them apart. So we have Shriner 1, Shriner 2, Shriner 3, and so on, for an infinite number of Shriners. We have one Shriner for each natural number.

Can we find a room in our motel for each Shriner? Of course! One easy way is to put Shriner 1 in Room 1, Shriner 2 in Room 2, Shriner 3 in Room 3, and so on. Each and every Shriner gets a room. Because we can match up the Shriners and the rooms this way, one-to-one, we say that these two infinite groups are "comparable." We will call infinite groups "equal" if they are comparable. What we are doing here is creating a new concept of "equal" that we can use with infinite groups.

Now back to our motel. The Shriners are about to check in, and our manager discovers she has a problem. The persons in Room 1 and Room 2 are enjoying the cable TV so much they refuse to leave. How is she going to find rooms for all the Shriners? Well, this is a snap for our mathematically gifted motel manager. She puts Shriner 1 in Room 3. She puts Shriner 2 in Room 4. She continues this process for all Shriners. She puts Shriner N in Room $N + 2$. Again, each and every Shriner gets a room! Infinity + 2 = infinity! (Remember, we are using the "=" sign for our new method of comparing infinite numbers.)

The Shriners are delighted with our motel, and they reserve it for next year's convention. But the night clerk makes a mistake. He accidentally books both the Shriners and the Masons (an equally infinite group) for the same night. Both groups show up. Can our manager find rooms for all of them?

You bet! Our mathematically gifted motel manager puts Shriner 1 in Room 1, and she puts Mason 1 in Room 2. She puts Shriner 2 in Room 3, and Mason 2 in Room 4. She continues like this and puts the Nth Shriner in Room $2N - 1$, and the Nth Mason in Room $2N$. Everyone gets a room! So using our new idea of comparing infinite groups, we say that the infinite number of Shriners is "equal" to the infinite number of rooms, which is "equal" to the infinite number of Shriners *plus* the infinite number of Masons! This seems truly strange. Infinity + infinity = infinity!

Well, the Shriners and the Masons are mighty pleased, and word spreads. In the third year, our manager gets calls not only from the Shriners and Masons but also from an infinite number of other infinite groups to book the very same night. Can she do it?

Again, this is no ordinary motel manager. She knows her year-end bonus could be fantastic if she can book an infinite number of infinite groups into the motel on the same night. But how? She starts by giving each group a number. (Let's just assume for now she can do this, but this is an important assumption that doesn't have to be true.) She calls the Shriners group 1, and the Masons group 2, and so on. She realizes she can assign to each person in each group his or her own unique pair of integers, of the form (n, m), which she defines to be the nth person in the mth group. With the person number first and the group number second, then person $(2, 1)$ is the second person in the first group (our friends the Shriners), which is the person we called Shriner 2 above. Person $(5, 2)$ is the fifth person in the second group, or the fifth Mason. Person (3, 4) is the third person in the fourth group. Now for every person there is one and only one pair of natural numbers, and for every pair of natural numbers there is one and only one person.

"So what?" you might say. Well, now you can put in a single order every person in every group. Start with the one person whose numbers add up to 2—that's person (1, 1), the first Shriner. Then take the two persons whose numbers add up to 3—that's person (2, 1) the second Shriner and person (1, 2) the first Mason. Then take the persons whose numbers add up to 4—persons (3, 1), (2, 2), and (1, 3). Then take the persons whose numbers add up to 5, and keep going. You can easily put all of these people in a single order, and now you can assign each a room! Our mathematically gifted motel manager has done it again! Infinity "times" infinity = infinity! An infinite number of infinite groups is comparable to a single infinite group!

A Greater Infinity

At this point, you may think there isn't much to this infinity business. Is the concept of infinity so elastic, so flexible, that all infinities are comparable? So far, the kinds of infinite groups we have been considering are all comparable to the natural numbers: 1, 2, 3, 4, and so on. We call this kind of infinite group "countable." So instead of saying generally that infinity times infinity = infinity, what we should say, to be more careful, is that a countable infinity of countable infinities is still a countable infinity.

Now let's consider a different kind of infinite group. Consider the group, or set, of all real numbers between 0 and 1. This set includes not just the fractions (numbers of the form a/b, where a and b are integers and b is not zero) but an infinite number of "irrational" numbers. Two examples of irrational numbers in this set are $\sqrt{2}/2$ (an "algebraic" irrational number) and $\pi/4$ (a "transcendental" irrational number).

In 1896, George Cantor proved that this group (set) is *not* comparable to the group (set) of natural numbers. He used an old but powerful mathematical tool, proof by contradiction. Each of these real numbers can be written as an infinite decimal, of the form $0.a_1 a_2 a_3 \ldots$, where the a_i's are digits from 0 to 9. The a_i's may or may not have a pattern to them. (It can be shown that the a_i's will have a repeating pattern at some point if and only if the number is a fraction.) Suppose this group of real numbers is countable. This would mean that there is a way to arrange the real numbers from 0 to 1 so that we could assign to each real number exactly one natural number. Remember, this is the definition of a "countable" infinity, that we can "count" them. Then we could make a list, like this:

0.473219 …

0.521687 …

0.845913 …

and so on. (These could be any numbers.) If there were such an arrangement of the real numbers from 0 to 1, we could create a real number between 0 and 1 that is not in the above list. How? Start by picking a number from 0 to 9 that is not equal to the first digit to the right of the decimal point in the first number (in other words, not equal to 4 using the above list). Then pick a digit not equal to the second digit to the right of the decimal point in the second number (in other words, not equal to 2 using the above list). Continue throughout the infinite list, picking for the ith number on the list an ith digit not equal to the ith digit of the ith number.

Now this new number we are creating is not equal to the first number on our list, because it has a different first digit. It is not equal to the second number on our list, because it has a different second digit from the second number. We can continue on, picking a third digit different from the third digit of the third number, and so on, throughout the entire list. Our new number is between 0 and 1, and yet it is not equal to any number on the list. This contradicts our assumption that the real numbers between 0 and 1 are countable, and we are done!

We have proved that not all infinities are comparable and that there are different kinds of infinities! In fact, there are an infinite number of different kinds of infinities, although the others don't seem to have much, if any, practical use.

In the 1930s, George Cantor's proof was used to show that there are statements in mathematics that cannot be proved or disproved. They cannot be proved either true or false! A clever way was found, using prime numbers, to show that any possible mathematical proof can be assigned a unique natural number. Thus, the number of possible mathematical proofs is countable. It was then shown that the number of possible mathematical statements one can make is infinite and not countable. The conclusion—that some mathematical statements cannot be proved or disproved—stunned many mathematicians.

One question that comes up is whether there is any kind of infinity "between" the natural numbers and the real numbers. Stated another way, is there a concept of infinity that is greater than the natural numbers but less than the real numbers? Many tried, without success, to answer this question. We now know this question cannot be proved either false or true. It is an example of a mathematical statement that can neither be proved nor disproved.

NOTES

Chapter 1 The Great Question

1. Alfred North Whitehead, "Religion and Science," *Atlantic*, August 1925.

2. This cultural war is documented by Ben Stein in *Expelled: No Intelligence Allowed*, directed by Nathan Frankowski (Vivendi Entertainment, 2008).

3. Timothy Keller, *The Reason for God: Belief in an Age of Skepticism* (New York: Riverhead Books, 2008), p. xviii.

Chapter 2 The Good News

1. Timothy Keller, *The Reason for God: Belief in an Age of Skepticism* (New York: Riverhead Books, 2008).

Chapter 4 Religion versus Scientism

1. http://www.thearda.com/QL2010/QuickList_125.asp. Despite government restrictions, Christianity is growing explosively in China.

2. Genesis 32:24–30.

3. William Dembski, *The Design Revolution* (Downers Grove, IL: InterVarsity Press, 2004), 21.

4. Stephen Hawking, during an interview with Ken Campbell on the 1995 show *Reality on the Rocks: Beyond Our Ken*, http://en.wikiquote.org/wiki/Stephen_Hawking. Here's a similar statement from a scientist trying to make us feel small: Human beings are just "carbon based bags of mostly water on a speck of iron-silicate dust revolving around a boring dwarf star in a minor

galaxy in an underpopulated local group of galaxies in an unfashionable suburb of a supercluster." This quote is contained in Jack Huberman, ed., *The Quotable Atheist* (New York: Nation Books, 2007), p. 313, and is attributed to Peter Walker, a space physicist at Rice University.

5. Steven Weinberg, *The First Three Minutes: A Modern View of the Origin of the Universe* (New York: Basic Books, 1979), p. 154.

6. Alvin Plantinga, *Where the Conflict Really Lies: Science, Religion, and Naturalism* (Oxford: Oxford University Press, 2012), quoted by Thomas Nagel, in "A Philosopher Defends Religion," *New York Times Review of Books*, September 27, 2012.

7. This is from one of his lectures; you can view the video at

   http://amiquote.tumblr.com/post/4463599197/richard-feynman- on-how-we-would-look-for-a-new-law.

8. http://www.gallup.com/poll/147887/americans-continue-believe-god.aspx.

Chapter 5 Paradigm Blindness

1. Jerzy Dobrzycki and Leszek Hajdukiewicz, "Kopernik, Mikołaj," in *Polski słownik biograficzny* [Polish biographical dictionary], vol. 14 (Krakow: Polish Academy of Sciences, 1969), p. 11.

2. Richard Panek, *The 4 Percent Universe: Dark Matter, Dark Energy, and the Race to Discover the Rest of Reality* (New York: Houghton Mifflin, 2011), p. 8.

3. In a 2005 survey, members of Britain's Royal Society voted that Newton's contributions to science and humanity were greater than those of Einstein. Royal Society, "Newton Beats Einstein in Polls of Royal Society Scientists and the Public," http://royalsociety.org/News.aspx?id=1324&terms=Newton+beats+Einstein+in+polls+of+scientists+and+the+public. See also Michael H. Hart, *The 100: A Ranking of the Most Influential Persons in History* (New York: Carol Publishing Group/Citadel Press, 1978; repr., 1992).

4. Letter to Robert Hooke, 15 February 1676.

5. S. Herrmann, A. Senger, K. Möhle, M. Nagel, E. V. Kovalchuk, and A. Peters, "Rotating Optical Cavity Experiment Testing Lorentz Invariance at the 10^{-17} Level," *Physical Review D* 80, no. 100 (2009): 105011.

6. Max Planck, *Scientific Autobiography and Other Papers*, trans. F. Gaynor (New York: Philosophical Library, 1949), pp. 33–34.

Chapter 6 The Great Debate

1. William Dembski, *The Design Revolution: Answering the Toughest Question about Intelligent Design* (Downers Grove, IL: InterVarsity Press, 2004), p. 21.

2. Richard Dawkins, *The Blind Watchmaker: Why the Evidence of Evolution Reveals a Universe Without Design* (New York: Norton, 1986).

3. *Kitzmiller v. Dover Area School District*, 400 F. Supp. 2d 707 (2005). The decision essentially recites the brief of the ACLU that Intelligent Design is not "science."

4. See http://www.bbc.co.uk/news/education-20547195.

5. William Dembski, *The Design Inference: Eliminating Chance through Small Probabilities* (Cambridge: Cambridge University Press, 1998), p. 27.

6. *Easton's Bible Dictionary* (Thomas Nelson, 1897).

7. 2 Chronicles 32:10, 13–14.

Chapter 7 Creation

1. R. H. Dicke, P. J. E. Peebles, P. J. Roll, and D. T. Wilkinson, "Cosmic Black-Body Radiation," *Astrophysical Journal* 142 (1965): 414–19.

2. David Schram, "Dark Matter and the Origin of Cosmic Structure," *Sky & Telescope* (October 1994): 29.

3. William Dembski, *The Design Inference: Eliminating Chance through Small Probabilities* (Cambridge: Cambridge University Press, 1998), p. 38.

Chapter 8 Fine-Tuning

1. http://hyperphysics.phy-astr.gsu.edu/hbase/astro/denpar.html.

2. Paul Davies, *The Accidental Universe* (Cambridge: Cambridge University Press, 1982).

3. Ibid., preface.

4. Ibid., p. 89.

5. Max Tegmark, Anthony Aguirre, Martin Rees, and Frank Wilczek, "Dimensionless Constants, Cosmology, and Other Dark Matter," *Physical Review Part D* (January 2006).

6. Freeman Dyson, *Disturbing the Universe* (New York: Harper & Row, 1979), p. 250.

7. Stephen Hawking, *A Brief History of Time* (New York: Bantam Books 1988), p. 125.

8. See "Why Some Scientist Embrace the 'Multiverse': This Universe's Evidence Suggests a Designing Intelligence, so Atheists Resort to an Idea with No Evidence," *National Review Online*, June 18, 2013, http://www.nationalreview.com/article/351319/why-some-scientists-embrace-multiverse-dennis-prager (accessed July 18, 2013).

9. Whitney Clavin and Alan Buis, "Astronomers Find Largest, Most Distant Reservoir of Water," NASA, July 22, 2011," http://www.nasa.gov/topics/universe/features/universe20110722.html (retrieved July 25, 2011); Staff, "Astronomers Find Largest, Oldest Mass of Water in Universe," http://www.space.com/ 12400-universe-biggest-oldest-cloud-water.html (retrieved July 23, 2011).

10. Michael J. Denton, *Nature's Destiny: How the Laws of Biology Reveal Purpose in the Universe* (New York: Free Press, 1998), p. 19. See, generally, chapter 2, "The Vital Fluid."

11. Ibid.

12. See http://www.snowballearth.org/.

13. Roger Penrose, *The Road to Reality: A Complete Guide to the Laws of the Universe* (London: Random House, 2004), pp. 726–32.

Chapter 9 Problems with the Multiverse

1. Martin Rees, *Just Six Numbers: The Deep Forces That Shape the Universe* (London: Weidenfeld & Nicolson, 1999).

Chapter 10 The Origin of Life

1. David Deutsch, *The Beginning of Infinity: Explanations That Transform the World* (New York: Viking Press, 2011), p. 143.

2. Michael J. Behe, *Darwin's Black Box: The Biochemical Challenge to Evolution* (New York: Free Press, 1996), p. 185.

3. See S. Iglesias-Groth, A. Manchado, R. Rebolo, J. I. Gonzalez Hernandez, D. A. Garcia-Hernandez, and D. L. Lambert, "A Search for Interstellar Anthracene toward the Perseus Anomalous Microwave Emission Region," *Monthly Notices of the Royal Astronomical Society* 407, no. 4 (May 2010): 2157; Jan Cami, Jeronimo Bernard-Salas, Els Peeters, and Sarah Elizabeth Malak,

"Detection of C60 and C70 in a Young Planetary Nebula," *Science* 329, no. 5996 (July 22, 2010): 1180–82. See also "Extraterrestrial Nucleobases in the Murchison Meteorite," *Earth and Planetary Science Letters* 270 (2008) 130–36, http://astrobiology.gsfc.nasa.gov/analytical/PDF/Martinsetal2008.pdf.

4. Casey Luskin, "Not Making the Grade: An Evaluation of 22 Recent Biology Textbooks and Their Use of Selection Icons of Evolution," September 26, 2011, http://www.evolutionnews.org/DiscoveryInstitute_2011TextbookReview.pdf.

5. John Cohen, "Novel Center Seeks to Add Spark to Origins of Life," *Science* 270 (1995): 1925–26.

6. Deborah Kelley, "Is It Time to Throw Out 'Primordial Soup' Theory?" NPR, February 7, 2010.

7. Committee on the Limits of Organic Life in Planetary Systems, Committee on the Origins and Evolution of Life, National Research Council, *The Limits of Organic Life in Planetary Systems* (Washington, DC: National Academy Press, 2007), p. 60.

8. Marcel P. Schutzenberger, "Mathematical Challenges to the Neo-Darwinian Interpretation of Evolution," *Proceedings of the Wistar Symposium* (Philadelphia: Wistar Institute Press, 1967), p. 73.

9. Fred Hoyle, *The Intelligent Universe: A New View of Creation and Evolution* (New York: Holt, Rinehart & Winston, 1983), p. 23.

10. Robert Shapiro, *Origins: A Skeptic's Guide to the Creation of Life on Earth* (New York: Summit Books, 1996), p. 116.

11. Harold C. Urey, quoted in *Christian Science Monitor* (January 4, 1962), 4.

12. Stanley L. Miller and H. James Cleaves, "Prebiotic Chemistry on the Primitive Earth," in *Systems Biology: Volume 1. Genomics*, ed. Isidore and Gregory Stephanopoulos (New York: Oxford University Press, 2007), p. 3.

13. See http://origins.harvard.edu/.

14. Anna Kushnir, "The Origins of Life on Earth. Really," March 9, 2009, http://blogs.nature.com/boston/2009/03/09/the-origins-of-life-on-earth-really.

15. Eugene V. Koonin, *The Logic of Chance: The Nature and Origin of Biological Evolution* (Upper Saddle River, NJ: FT Press, 2011), p. 391.

16. "Rover Finds Hint of Life's Cradle on Mars," *New Scientist*, March 23–29, 2013, p. 8.

17. For the NASA website, see http://history.nasa.gov/SP-4212/ch11-5.html (last accessed September 30, 2013). Dr. Gil Levin's website is at http://www.gillevin.com/mars.htm (last accessed September 30, 2013).

18. See also Stephen Webb, *If the Universe Is Teeming with Aliens . . . Where Is Everybody? Fifty Solutions to Fermi's Paradox and the Problem of Extraterrestrial Life* (New York: Copernicus Books, 2002).

19. See http://www.britannica.com/EBchecked/topic/560859/spontaneous-generation.

20. Colin Munn, *Marine Microbiology: Ecology and Applications*, 2nd ed. (New York: Garland Science, 2011).

21. Naoki Sato, "Comparative Analysis of the Genomes of Cyanobacteria and Plants," *Genome Informatics* 13 (2002): 173–82.

22. See John Reidhaar-Olson and Robert Sauer," Functionally Acceptable Substitutions in Two Alpha-Helical Regions of Lambda Repressor," *Proteins: Structure, Function, and Genetics* 7, no. 4 (1990): 306–16; James Bowie and Robert Sauer, "Identifying the Determinants of Folding and Activity for a Protein of Unknown Structure," *Proceedings of the National Academy of Sciences USA* 86 (1989): 2152–56.

23. Douglas Axe, "Estimating the Prevalence of Protein Sequences Adopting Functional Enzyme Folds," *Journal of Molecular Biology* 341, no. 5 (August 24 2004): 1295–1315.

24. A. G. Cairns-Smith, *Seven Clues to the Origin of Life: A Scientific Detective Story* (Cambridge: Cambridge University Press, 1985; repr., 1993), pp. 46–47.

25. Fred Hoyle, *New Scientist*, November 1981. Rabbi Moshe Averick paraphrased the end of this paragraph for the title of his book, *Nonsense of a High Order: The Confused and Illusionary World of the Atheist* (Chicago: Tradition & Reason Press, 2010).

26. http://www.evolutionnews.org/2011/06/life_purpose_mind_where_the_ma046991.html.

27. Harold Morowitz, *Energy Flow in Biology* (New York: Academic Press, 1968).

Chapter 11 The Technology of Life

1. Michael Denton, *Evolution: A Theory in Crisis* (Chevy Chase, MD: Adler & Adler, 1986), p. 328. There are videos of how cells work on YouTube, go to https://www.youtube.com/user/discoveryinstitute. I recommend them.

But be careful. The Discovery Institute is a lightning rod for the ire of every die-hard Darwinist in the universe, and if you search on YouTube for Discovery Institute, you will also see their attacks posted.

2. See http://www.youtube.com/watch?v=D720rzuIuv8 (accessed September 2, 2013). Some believe that an organism can rewrite its own DNA code, that the DNA code is not just "read only." See James Shapiro, "How Life Changes Itself: The Read-Write (RW) Genome," *Physics of Life Reviews*, July 8, 2013, http://www.ncbi.nlm. nih.gov/pubmed/23876611 (accessed September 1, 2013). See also A. Slack, P. C. Thornton, D. B. Magner, S. M. Rosenberg, and P. J. Hastings, "On the Mechanism of Gene Amplification Induced under Stress in *Escherichia coli*," *PLoS Genetics* 2, no. 4 (2006): 385–398; Megan N. Hersh, Rebecca G. Ponder, P. J. Hastings, and Susan M. Rosenberg, "Adaptive Mutation and Amplification in *Escherichia coli*: Two Pathways of Genome Adaptation under Stress," *Research in Microbiology* 155 (2004): 352–59.

3. Chapter 14 of Stephen Meyer's *Darwin's Doubt* (New York: HarperOne, 2013) contains a good summary of this area.

4. Bruce Alberts, Alexander Johnson, Julian Lewis, Martin Raff, Keith Roberts, and Peter Walter, *Molecular Biology of the Cell*, 5th ed. (New York: Garland Science, 2008), p. 2.

5. Stephen P. Meyer, *Signature in the Cell: DNA and the Evidence for Intelligent Design* (New York: HarperOne, 2009). See chapter 14, "The RNA World," beginning on page 296.

6. Michael J. Denton, *Nature's Destiny: How the Laws of Biology Reveal Purpose in the Universe* (New York: Free Press, 1998), p. 141. See, generally, chapter 7, "The Double Helix."

7. Available at http://www.extremetech.com/extreme/134672-harvard-cracks-dna-storage-crams-700-terabytes-of-data-into-a-single-gram.

8. Denton, *Evolution: A Theory in Crisis*, p. 334.

9. See "Double Helix Serves Double Duty," *New York Times*, January 29, 2013, Science section, p. D3.

10. Alberts et al., *Molecular Biology of the Cell*, p. 63.

11. M. S. Jurica and M. J. Moore, "Pre-mRNA Splicing: Awash in a Sea of Proteins," *Molecular Cell* 12 (2003): 5–14.

12. The wonders of the spliceosome go on and on. See http://crev.info/ 2013/06/ wonders-of-the-spliceosome-coming-to-light/ (accessed July17, 2013).

13. Here's a nice video about the spliceosome, http://www.youtube.com/watch?v=U_5yJYRvh8A (accessed August 31, 2013).

14. Alberts et al., *Molecular Biology of the Cell*, p. 125.

15. Ibid., pp. 6–7

16. G. K. Philip and S. J. Freeland, "Did Evolution Select a Nonrandom 'Alphabet' of Amino Acids?" *Astrobiology* 11, no. 3 (April 2011): 235–40.

17. See Jonathan McLatchie, "The Finely Tuned Genetic Code," *Evolution News and Views*, November 19, 2011, http://www.evolutionnews.org/2011/11/the_finely_tuned_genetic_code052611.html.

18. Stephen J. Freeland, Robin D. Knight, Laura F. Landweber, and Laurence D. Hurst, "Early Fixation of an Optimal Genetic Code," *Molecular Biology and Evolution* 17, no. 4 (2000): 511–18, http://mbe.oxfordjournals.org/content/17/4/511.full.

19. E. V. Koonin and A. S. Novozhilov, "Origin and Evolution of the Genetic Code: The Universal Enigma," *IUBMB Life* 61, no. 2 (2009): 99–111, abstract available at http://www.ncbi.nlm.nih.gov/pubmed/19117371 (accessed August 31, 2013).

20. Some bacteria produce right-handed amino acids to use in cell walls to slow growth. "Right-Handed Amino Acids Help Bacteria Adapt," Howard Hughes Medical Institute. http://www.hhmi.org/news/right-handed-amino-acids-help-bacteria-adapt (accessed August 28, 2013).

21. Alberts et al., *Molecular Biology of the Cell*, pp. 130, 137.

22. http://www.sciencedaily.com/releases/2013/02/130204094606.htm.

23. Alberts et al., *Molecular Biology of the Cell*, pp. 264, 280.

24. Bruce Alberts, Alexander Johnson, Julian Lewis, Martin Raff, Keith Roberts, and Peter Walter, *Molecular Biology of the Cell*, 4th ed. (New York: Garland Science, 2002), p. 237.

25. Shapiro, "How Life Changes Itself."

26. See http://www.evolutionnews.org/2013/02/some_perspectiv068951.html.

27. http://www.ncbi.nlm.nih.gov/books/NBK9841/.

28. Alberts et al., *Molecular Biology of the Cell*, 5th ed., p. 296.

29. Ibid., p. 295.

30. Ibid., pp. 211, 220.

Chapter 12 Puzzles of Macroevolution

1. Moshe Averick, *Nonsense of a High Order: The Confused and Illusory World of the Atheist* (Chicago: Tradition & Reason Press, 2010).

2. See Michael A. Flannery, *Alfred Russel Wallace: A Rediscovered Life* (Seattle: Discovery Institute Press, 2011).

3. See, generally, Jonathan Weiner, *The Beak of the Finch* (New York: Vintage Books, 1994); Jeffrey Podos and Stephen Nowicki, "Beaks, Adaptation, and Vocal Evolution in Darwin's Finches," *BioScience* 54, no. 6 (June 2004): 501–10; Peter R. Grant and B. Rosemary Grant, "Predicting Microevolutionary Responses to Directional Selection on Heritable Variation," *Evolution* 49 (1995): 241–51; Peter R. Grant and B. Rosemary Grant, "Speciation and Hybridization of Birds on Islands," in *Evolution on Islands*, ed. Peter R. Grant (Oxford: Oxford University Press, 1998), pp. 142–62.

4. Bruce Alberts, Alexander Johnson, Julian Lewis, Martin Raff, Keith Roberts, and Peter Walter, *Molecular Biology of the Cell*, 5th ed. (New York: Garland Science, 2008), p. 1.

5. See "All Alone," *New Scientist* 217, no. 2900 (January 19, 2013): 40–43.

6. J. W. Schopf and B. M. Packer, "Early Archean (3.5-Billion- to 3.3-Billion-Year-Old) Microfossils from Warrawoona Group, Australia," *Science* 237, no. 4810 (1987): 70; J. W. Schopf, "Microfossils of the Early Archean Apex Chert," *Science* 260 (April 30, 1993).

7. See Stephen Meyer, *Darwin's Doubt* (New York: HarperOne, 2013), p. 82.

8. Charles Darwin, *The Origin of Species* (1859; repr., London: Penguin, 1985), p. 302.

9. Douglas Erwin, Marc Laflamme, Sarah Tweedt, Erik Sperling, Davide Pisani, and Kevin Peterson, "The Cambrian Conundrum: Early Divergence and Later Ecological Success in the Early History of Animals," *Science* 334 (November 25, 2011): 1091–97.

10. Ibid, table S-2.

11. Meyer, *Darwin's Doubt*, p. 73. For a detailed discussion of the conclusions of various experts of the time periods involved and the suddenness of the explosion, see Casey Luskin, "How 'Sudden' Was the Cambrian Explosion?" *Evolution News and Views*, July 16, 2013, http://www.evolutionnews.org/2013/07/how_sudden_was_074511.html (accessed July 17, 2013).

12. Jeffrey H. Schwartz, *Sudden Origins: Fossils, Genes, and the Emergence of Species* (New York: Wiley, 1999), p. 3.

13. Richard M. Bateman, Peter R. Crane, William A. DiMichele, Paul R. Kenrick, Nick P. Rowe, Thomas Speck, and William E. Stein, "Early Evolution of Land Plants: Phylogeny, Physiology, and Ecology of the Primary Terrestrial Radiation," *Annual Review of Ecology and Systematics* 29 (1998): 263–92.

14. Arthur Strahler, *Science and Earth History: The Evolution/Creation Controversy* (Amherst, NY: Prometheus Books, 1987), p. 408.

15. A. Cooper and R. Fortey, "Evolutionary Explosions and the Phylogenetic Fuse," *Trends in Ecology and Evolution* 13, no. 4 (1998): 151–56.

16. Darwin, *The Origin of Species,* p. 292.

17. Steven M. Stanley, *Macroevolution: Pattern and Process* (New York: W. H. Freeman, 1979), p. 39.

18. Quoted in Alexander Mebane, *Darwin's Creation-Myth* (Venice, FL: P&D Printing, 1994), p. 18.

19. Stephen Jay Gould, "Is a New and General Theory of Evolution Emerging?" *Paleobiology* 6, no. 1 (January 1980): 127.

20. Robert A. Martin, *Missing Links: Evolutionary Concepts & Transitions Through Time* (Sudbury, MA: Jones and Bartlett, 2004), p. 153; Carl C. Swisher III, Yuan-qing Wang, Xiao-lin Wang, Xing Xu, and Yuan Wang, "Cretaceous Age for the Feathered Dinosaurs of Lianoing, China," *Nature* 400 (July 1, 1999): 58–61; Alan Feduccia, *The Origin and Evolution of Birds*, 2nd ed. (New Haven, CT: Yale University Press, 1999); Devon E. Quick and John A. Ruben, "Cardio-Pulmonary Anatomy in Theropod Dinosaurs: Implications from Extant Archosaurs," *Journal of Morphology* 270 (2009): 1232–46; Frances C. James and John A. Pourtless IV, "Cladistics and the Origins of Birds: A Review and Two New Analyses," *Ornithological Monographs* 66 (2009): 1–78.

21. Stephen Jay Gould, "The Return of Hopeful Monsters," *Natural History* 86, no. 6 (June–July 1977): 24.

22. Fred Hoyle, *The Mathematics of Evolution* (Acorn Enterprises LLC, 1999); originally available in manuscript facsimile as *Mathematics of Evolution*, Weston Publication on the Cosmic Origin of Life No. 1, 1987. In the preface he states: "The criticism of the Darwinian theory given in this book arises straightforwardly from my belief that the theory is wrong."

23. David Berlinski, "The Deniable Darwin," *Commentary* 101, no. 6 (June 1, 1996), http://www.discovery.org/a/130.

24. A. K. Gauger and D. D. Axe, "The Evolutionary Accessibility of New Enzyme Functions: A Case Study from the Biotin Pathway," *BIO-Complexity* 2, no. 1 (2011): 1–17.

25. http://www.miamiherald.com/2002/12/04/928815/building-takes-your-breath-away.html.

26. http://www.arn.org/docs/berlinski/db_deniabledarwin0696.htm.

27. Thomas H. Frazzetta, *Complex Adaptations in Evolving Populations* (Stamford, CT: Sinauer Associates, 1975), p. 20.

28. W. Wayt Gibbs, "The Unseen Genome: Gems among the Junk," *Scientific American*, November 2003 (emphasis added), http://www.scientificamerican.com/article.cfm?id=the-unseen-genome-gems-am.

29. Helen Pearson, "Genetic Information Codes and Enigmas," *Nature* 444, no. 259 (November 16, 2006). See also Casey Luskin, "Junk DNA and Science-Stopping," http://www.evolutionnews.org/2006/12/junk_dna_and_sciencestopping002886.html.

30. The ENCODE Project Consortium, "An Integrated Encyclopedia of DNA Elements in the Human Genome," *Nature* 489, no. 7414 (2012): 57–74.

31. E. Yong, "ENCODE: The Rough Guide to the Human Genome," *Discover Magazine*, September 5, 2012.

32. *New York Times*, September 6, 2012, p. 1.

33. William A. Dembski, "Science and Design," *First Things* 86 (October 1, 1998).

34. Richard Dawkins, *The Greatest Show on Earth: The Evidence for Evolution* (New York: Free Press, 2009), pp. 332–33.

35. See "Rare Japanese Plant Has Largest Genome Known to Science," *Science Daily*, October 7, 2010, http://www.sciencedaily.com/releases/2010/10/101007120641.htm.

36. See http://genome.ucsc.edu/ENCODE/ (accessed July 20, 2013).

37. Christopher Vollmers, Robert J. Schmitz, Jason Nathanson, Gene Yeo, Joseph R. Ecker, and Satchidananda Panda, "Circadian Oscillations of Protein-Coding and Regulatory RNAs in a Highly Dynamic Mammalian Liver Epigenome," *Cell Metabolism* 17, no. 6 (December 2012): 833–45.

38. Meyer, *Darwin's Doubt*, p. 206.

39. John F. Reidhaar-Olson and Robert T. Sauer, "Functionally Acceptable Substitutions in Two α-Helical Regions of a λ Repressor," *Proteins: Structure, Function, and Genetics* 7 (1990): 315.

40. http://creation.com/germ-7-motors-in-1.

41. Geoffrey Simmons, MD, *Billions of Missing Links: A Rational Look at the Mysteries Evolution Can't Explain* (Eugene, OR: Harvest Hour Publishers, 2007).

42. B. W. Bowen, F. A. Abreu-Grobois, G. H. Balazs, N. Kamezaki, C. J. Limpus, and R. J. Ferl, "Trans-Pacific Migrations of the Loggerhead Turtle (*Caretta caretta*) Demonstrated with Mitochondrial Markers," *Proceedings of the National Academy of Sciences* 92, no. 9 (1995): 3731–34.

43. See, e.g., http://www.livescience.com/21080-loggerhead-turtle-migration.html.

44. "Animal Magnetism: First Evidence That Magnetism Helps Salmon Find Home," *Science News*, February 7, 2013, http://www.sciencedaily.com/releases/2013/02/130207131713.htm.

45. See "All Alone," pp. 40–43.

46. Ibid., p. 41.

47. K. Khalturin, G. Hemmrich, S. Fraune, R. Augistin, and T. C. Bosch, "More Than Just Orphans: Are Taxonomically-Restricted Genes Important in Evolution?" *Trends in Genetics* 25, no. 9 (2009): 404–13.

48. Garret Surn et al., "The Genome Sequence of the Leaf-Cutter Ant *Atta cephalotex* Reveals Insights into Its Obligate Symbiotic Lifestyle," *PLoS Genetics* 7 (2011): e1002007.

49. Daniel F. Simola et al., "Social Insect Genomes Exhibit Dramatic Evolution in Gene Composition and Regulation While Preserving Regulatory Features Linked to Sociality." *Genome Research* 23, no. 8 (2013): 1235–47.

50. Ann Gauger, Douglas Axe, and Casey Luskin, *Science and Human Origins* (Seattle: Discovery Institute Press, 2012), p. 18.

51. Alberts et al., *Molecular Biology of the Cell*, p. 206. The mean length of each "piece" of spliced DNA (called an "exon") is 145 letters.

52. "Relative Differences: The Myth of 1%," *Science* 316 (June 29, 2007): 1836.

53. Gauger, Axe, and Luskin, *Science and Human Origins*, p. 21.

54. Dennis M. Bramble and Daniel E. Lieberman, "Endurance Running and the Evolution of Homo," *Nature* 432 (2004): 345–52.

55. Gauger, Axe, and Luskin, *Science and Human Origins*, 43; J. P. Demuth, T. De Bie, J. E. Stajich, N. Cristianini, and M. W. Hahn, "The Evolution of Mammalian Gene Families," *PLoS One* 1 (2006): e85.

56. "All Alone," p. 44.

57. See "Pretty Useful: Appendix Evolved More Than 30 Times," *Science Now*, February 13, 2013, http://www.wired.com/wiredscience/2013/02/appendix-revolution/?cid=co5811204.

58. Joe Parker, Georgia Tsagkogeorga, James A. Cotton, Yuan Liu, Paolo Provero, Elia Stupka, and Stephen J. Rossiter, "Genome-Wide Signatures of Convergent Evolution in Echolocating Mammals," *Nature* (2013), http://www.nature.com/nature/journal/vaop/ncurrent/full/nature12511.html?WT.ec_id=NATURE-20130905 (accessed September 6, 2013).

59. Lad Allen, dir., *Flight: The Genius of Birds*, DVD (Illustra Media, May 2013).

Chapter 13 Our Special Earth

1. See http://www.nasa.gov/mission_pages/kepler/news/kepler20130103.html; http://www.bbc.co.uk/news/science-environment-20942440. The "17 billion" number comes from assuming 100 billion stars in our galaxy, which is probably too low.

2. Guillermo Gonzalez and Jay Richards, *The Privileged Planet: How Our Place in the Cosmos Is Designed for Discovery* (Washington, DC: Regnery, 2004), p. 341.

3. Michael Rowan-Robinson, *Cosmology*, 3rd ed. (Oxford: Clarendon Press, 1996), p. 62.

4. Peter D. Ward and Donald Brownlee, *Rare Earth: Why Complex Life Is Uncommon in the Universe* (New York: Copernicus Books, 2004), p. 29.

5. Ibid., p. xxxii.

6. Gonzalez and Richards, *Privileged Planet*, p. 167.

7. Martin Beech, "The Past, Present and Future Threat to Earth's Biosphere," *Astrophysics and Space Science* 336 (2011): 287–302; A. L. Melott and B. C. Thomas, "Astrophysical Ionizing Radiation and Earth: A Brief Review and Census of Intermittent Sources," *Astrobiology* 11 (2011): 343–61.

8. Melott and Thomas, "Astrophysical Ionizing Radiation and Earth."

9. Narciso Benitez, Jesus Maiz-Apellaniz, and Matilde Canelles, "Evidence for Nearby Supernova Explosions," *Physical Review Letters* 88, no. 8 (2002): 081101.

10. Charles Lineweaver, Yeshe Fenner, and Brad K. Gibson, "The Galactic Habitable Zone and the Age Distribution of Complex Life in the Milky Way," *Science* 303, no. 5654 (2004): 59–62.

11. See, generally, http://en.wikipedia.org/wiki/Tunguska_event (accessed August 23, 2013).

12. E. F. Tedesco and F.-X. Desert, "The Infrared Space Observatory Deep Asteroid Search," *Astronomical Journal* 123, no. 4 (2002): 2070–82.

13. NASA Science, "Ten-Thousandth Near-Earth Object Discovered," http://science.nasa.gov/science-news/science-at-nasa/2013/24jun_neo/ (accessed June 25, 2013).

14. Audrey Delsanti and David Jewitt, "The Solar System Beyond The Planets," Institute for Astronomy, University of Hawaii, archived from the original on September 25, 2007; see also G. A. Krasinsky, E. V. Pitjeva, M. V. Vasilyev, and E. I. Yagudina, "Hidden Mass in the Asteroid Belt," *Icarus* 158, no. 1 (July 2002): 98–105.

15. V. V. Emelyanenko, D. J. Asher, and M. E. Bailey, "The Fundamental Role of the Oort Cloud in Determining the Flux of Comets through the Planetary System," *Monthly Notices of the Royal Astronomical Society* 381, no. 2 (2007): 779–89.

16. Gonzalez and Richards, *Privileged Planet*, p. 74.

17. Ward and Brownlee, *Rare Earth*, pp. 22–23.

18. Ibid., p. 24.

19. G. W. Lockwood, B. A. Skiff, and R. R. Radick, "The Photometric Variability of Sun-like Stars: Observations and Results, 1984–1995," *Astrophysical Journal* 485 (1997): 789–811.

20. Juan J. Jimenez-Torres, Barbara Pichardo, George Lake, and Henry Throop, "Effect of Different Stellar Galactic Environments on Planetary Discs: I. The Solar Neighbourhood and the Birth Cloud of the Sun," *Monthly Notices of the Royal Astronomical Society* 418 (2013): 1272–84.

21. J. Horner, B. W. Jones, and J. Chambers, "Jupiter Friend or Foe? III. The Oort Cloud Comets," *International Journal of Astrobiology* 9, no. 1 (2010): 1–10.

22. R. Gomes, H. F. Levison, K. Tsiganis, and A. Mrbidelli, "Origin of the Cataclysmic Late Heavy Bombardment Period of the Terrestrial Planets," *Nature* 435, no. 7041 (2005): 466–69.

23. Rebecca G. Martin and Mario Livio, "On the Formation and Evolution of Asteroid Belts and Their Potential Significance for Life," *Monthly Notices of the Royal Astronomical Society* 428 (2013): L11–L15.

24. See "Alien Life May Require Rare 'Just-Right' Asteroid Belts," Space.com, November 2, 2012, http://www.space.com/18326-asteroid-belt-evolution-alien-life.html; Claire Moskowitz, "Solar Systems Like Ours May Be Rare," Space. com, July 21, 2008, http://www.space.com/5638-solar-systems-rare.html; Jeff Hecht, "Solar Systems Like Our May Be Rare," *New Scientist*, August 7, 2008, http://www.newscientist.com/article/dn14492-solar-systems-like-ours-may-be-rare.html.

25. R. K. Kopparapu et al., "Habitable Zones around Main-Sequence Stars: New Estimates," *Astrophysics Journal* 765, no. 2 (2013): 131.

26. Ron Cowen, "Common Source for Earth and Moon Water," *Nature News*, May 9, 2013, http://www.nature.com/news/common-source-for-earth-and-moon-water-1.12963 (accessed June 27, 2013).

27. "The Day the Earth Exploded," *New Scientist*, July 6–12, 2013, pp. 30–33.

28. http://www.cosmosmagazine.com/news/earths-moon-a-rare-species/.

29. Gonzalez and Richards, *Privileged Planet*, p. 341.

30. Ward and Brownlee, *Rare Earth*, p. 46.

31. See http://quake.mit.edu/hilstgroup/CoreMantle/EarthCompo.pdf.

32. Ward and Brownlee, *Rare Earth*, p. 46.

33. J. S. Lewis, *Worlds Without End: The Exploration of Planets Known and Unknown* (Reading, UK: Helix Books, 1998), p. 199.

34. Ward and Brownlee, *Rare Earth*, p. 220.

Chapter 14 A Foundation of Thought

1. Roger Penrose, *The Road to Reality: A Complete Guide to the Laws of the Universe* (New York: Knopf, 2005), pp. 1033–34. See also Mark Steiner, *The Applicability of Mathematics as a Philosophical Problem* (Cambridge, MA: Harvard University Press, 1998).

2. See Antoine Suarez and Peter Adams, *Is Science Compatible with Free Will? Exploring Free Will and Consciousness in the Light of Quantum Physics and Neuroscience* (New York: Springer, 2013).

3. "Quantum Wonders: Corpuscles and Buckyballs," *New Scientist*, May 6, 2010.

4. http://en.wikipedia.org/wiki/Quantum_entanglement. See also http://www.sciencedaily.com/articles/q/quantum_entanglement.htm.

5. See http://www.forbes.com/sites/alexknapp/2012/09/06/physicists-quantum-teleport-photons-over-88-miles/.

6. Matthew Francis, "Quantum Entanglement Shows That Reality Can't Be Local," *Ars Technica*, October 30, 2012.

7. "Reality Check," *New Scientist*, August 3–9, 2013, p. 34.

8. C. Brukner and A. Zeilinger, "Young's Experiment and the Finiteness of Information," *Philosophical Transactions of the Royal Society of London* 360 (2001): 1061–69.

9. Review of *Quantum Enigma*, by Richard Conn Henry, for the *Journal of Scientific Exploration* 21 (2006): 185.

10. Ibid. (emphasis in original).

11. Ibid. (emphasis in original).

12. John 1:1.

13. See http://www.quantumphil.org/index.htm (accessed September 2, 2013).

Chapter 15 The Logic of Belief

1. Dietrich Bonhoeffer, letter to Eberhard Bethge, May 29, 1944, in *Letters and Papers from Prison*, ed. Eberhard Bethge, trans. Reginald H. Fuller (New York: Touchstone, 1997), pp. 310–12; Translation of *Widerstand und Ergebung* (Munich: Christian Kaiser Verlag, 1970).

2. The ENCODE Project Consortium, "An Integrated Encyclopedia of DNA Elements in the Human Genome," *Nature* 489 (September 6, 2012): 57–74.

Chapter 16 Connecting the Dots

1. http://www.oldearth.org/yom_hebrew.htm.

2. John Mark Reynolds, "Thoughts on Faith and Science" (Part I), http://www.patheos.com/blogs/eidos/2013/07/thoughts-on-faith-and-science-part-i/ (accessed July 17, 2013).

3. Robert Jastrow, *God and the Astronomers*, 2nd ed. (New York: Reader's Library, 2000).

4. "Memories of Near Death Experiences: More Real Than Reality?" *Science Daily*, March 27, 2013, http://www.sciencedaily.com/releases/2013/03/130327190359.htm (accessed September 7, 2013).

5. Mary C. Neal, MD, *To Heaven and Back* (Colorado Springs, CO: WaterBrook Press 2011), p. 138.

6. Jeffrey Long, MD, with Paul Perry, *Evidence of the Afterlife: The Science of Near-Death Experiences* (New York: HarperOne 2010).

7. Ibid., p. 2.

8. The EEG "flatlines" in 10 to 20 seconds following cardiac arrest. See ibid.

9. Ibid. p. 64.

10. Ibid., p. 4 (emphasis in original).

11. For examples, see Robert Matthews, "Patients Near Death See Visions of Hell," *Daily Telegraph*, http://www.theforbiddenknowledge.com/hardtruth/visions_of_hell.htm.

12. http://www.near-death.com/storm.html.

13. John 14:6, New International Version, 1984.

14. John 14:9, New International Version, 1984.

15. John 14:10 New International Version, 1984.

16. C. S. Lewis, *The Problem of Pain* (London: Macmillan, 1961), p. 69.

BIBLIOGRAPHY

Alberts, Bruce, Alexander Johnson, Julian Lewis, Martin Raff, Keith Roberts, and Peter Walter. *Molecular Biology of the Cell*, 4th ed. New York: Garland Science, 2002.

———. *Molecular Biology of the Cell*, 5th ed. New York: Garland Science, 2008.

"Alien Life May Require Rare 'Just-Right' Asteroid Belts. Space.com, November 2, 2012. http://www.space.com/18326-asteroid-belt-evolution-alien-life.html.

"All Alone." *New Scientist* 217, no. 2900 (January 19, 2013): 40–43.

"Animal Magnetism: First Evidence That Magnetism Helps Salmon Find Home." *Science News*, February 7, 2013. http://www.sciencedaily.com/releases/2013/02/130207131713.htm.

Aron, Jacob. "Largest Structure Challenges Einstein's Smooth Cosmos." New Scientist, January 11, 2013. http://www.newscientist.com/article/dn23074-largest-structure-challenges-einsteins-smooth-cosmos.html.

"Astronomers Find Largest, Oldest Mass of Water in Universe." http://www.space.com/12400-universe-biggest-oldest-cloud-water.html. Accessed July 23, 2011.

Averick, Moshe. *Nonsense of a High Order: The Confused and Illusory World of the Atheist*. Chicago: Tradition & Reason Press, 2010.

Axe, Douglas. "Estimating the Prevalence of Protein Sequences Adopting Functional Enzyme Folds." *Journal of Molecular Biology* 341, no. 5 (August 24 2004): 1295–1315.

Bateman, Richard M., Peter R. Crane, William A. DiMichele, Paul R. Kenrick, Nick P. Rowe, Thomas Speck, and William E. Stein. "Early Evolution of Land Plants: Phylogeny, Physiology, and Ecology of the Primary Terrestrial Radiation." *Annual Review of Ecology and Systematics* 29 (1998): 263–92.

Beech, Martin. "The Past, Present and Future Threat to Earth's Biosphere." *Astrophysics and Space Science* 336 (2011): 287–302.

Behe, Michael J. *Darwin's Black Box: The Biochemical Challenge to Evolution*. New York: Free Press, 1996.

Benitez, Narciso, Jesus Maiz-Apellaniz, and Matilde Canelles. "Evidence for Nearby Supernova Explosions." *Physical Review Letters* 88, no. 8 (2002): 081101.

Benson, M. J., M. A. Wills, and R. Hitchin. "Quality of the Fossil Record Through Time," *Nature* 403 (February 3, 2000): 534–36.

Berlinski, David. "The Deniable Darwin." *Commentary* 101, no. 6 (June 1, 1996). http://www.discovery.org/a/130.

Bonhoeffer, Dietrich. Letter to Eberhard Bethge, 29 May 1944. In *Letters and Papers from Prison*, edited by Eberhard Bethge, translated by Reginald H. Fuller, pp. 310–12. New York: Touchstone, 1997; translation of *Widerstand und Ergebung* Munich: Christian Kaiser Verlag, 1970.

Bowen, B. W., F. A. Abreu-Grobois, G. H. Balazs, N. Kamezaki, C. J. Limpus, and R. J. Ferl. "Trans-Pacific Migrations of the Loggerhead Turtle (*Caretta caretta*) Demonstrated with Mitochondrial Markers." *Proceedings of the National Academy of Sciences* 92, no. 9 (1995): 3731–34.

Bowie, James, and Robert Sauer. "Identifying the Determinants of Folding and Activity for a Protein of Unknown Structure." *Proceedings of the National Academy of Sciences USA* 86 (1989): 2152–56.

Bramble, Dennis M., and Daniel E. Lieberman. "Endurance Running and the Evolution of Homo." *Nature* 432 (2004): 345–52.

Brukner, C., and A. Zeilinger. "Young's Experiment and the Finiteness of Information." *Philosophical Transactions of the Royal Society of London* 360 (2001): 1061–69.

Cairns-Smith, A. G. *Seven Clues to the Origin of Life: A Scientific Detective Story.* Cambridge: Cambridge University Press, 1985; repr., 1993.

Cami, Jeronimo Bernard-Salas, Els Peeters, and Sarah Elizabeth Malak. "Detection of C60 and C70 in a Young Planetary Nebula." *Science* 329, no. 5996 (July 22, 2010): 1180–82.

Clavin, Whitney, and Alan Buis. "Astronomers Find Largest, Most Distant Reservoir of Water." NASA, July 22, 2011. http://www.nasa.gov/topics/universe/features/universe20110722.html. Accessed July 25, 2011.

Cohen, John. "Novel Center Seeks to Add Spark to Origins of Life." *Science* 270 (1995): 1925–26.

Committee on the Limits of Organic Life in Planetary Systems, Committee on the Origins and Evolution of Life, National Research Council. *The Limits of Organic Life in Planetary Systems.* Washington, DC: National Academy Press, 2007.

"Common Source for Earth and Moon Water." *Nature News*, May 9, 2013. http://www.nature.com/news/common-source-for-earth-and-moon-water-1.12963. Accessed June 27, 2013.

Cooper, A., and R. Fortey. "Evolutionary Explosions and the Phylogenetic Fuse." *Trends in Ecology and Evolution* 13, no. 4 (1998): 151–56.

Cooper, Geoffrey M. *The Cell: A Molecular Approach*, 2nd ed. Sunderland, MA: Sinhauer, 2000.

Crick, Francis. "Central Dogma of Molecular Biology." *Nature* 227, no. 5258 (August 1970): 561–63.

Darwin, Charles. *The Origin of Species.* 1859; repr., London: Penguin, 1985.

Davies, Paul. *The Accidental Universe.* Cambridge: Cambridge University Press, 1982.

Dawkins, Richard D. *The Blind Watchmaker: Why the Evidence of Evolution Reveals a Universe Without Design.* New York: Norton, 1986.

———. *The Greatest Show on Earth: The Evidence for Evolution.* New York: Free Press, 2009.

"The Day the Earth Exploded." *New Scientist,* July 6–12, 2013, pp. 30–33.

Delsanti, Audrey, and David Jewitt. "The Solar System Beyond the Planets." Institute for Astronomy, University of Hawaii, archived from the original on September 25, 2007. http://web.archive.org/web/20070925203400/; http://www.ifa.hawaii.edu/faculty/jewitt/papers/20.06/DJ06.pdf. Accessed March 9, 2007.

Demski, William. *The Design Inference: Eliminating Chance through Small Probabilities.* Cambridge: Cambridge University Press, 1998.

———. *The Design Revolution: Answering the Toughest Question about Intelligent Design.* Downers Grove, IL: InterVarsity Press, 2004.

———. "Science and Design." *First Things* 86 (October 1, 1998).

Demuth, J. P., T. De Bie, J. E. Stajich, N. Cristianini, and M. W. Hahn. "The Evolution of Mammalian Gene Families." *PLoS One* 1 (2006): e85.

Denton, Michael. *Evolution: A Theory in Crisis.* Chevy Chase, MD: Adler & Adler, 1986.

———. *Nature's Destiny: How the Laws of Biology Reveal Purpose in the Universe.* New York: Free Press, 1998.

Deutsch, David. *The Beginning of Infinity: Explanations That Transform the World.* New York: Viking Press, 2011.

Dicke, R. H., P. J. E. Peebles, P. J. Roll, and D. T. Wilkinson. "Cosmic Black-Body Radiation." *Astrophysical Journal* 142 (1965): 414–19.

Dobrzycki, Jerzy, and Leszek Hajdukiewicz. "Kopernik, Mikołaj." In *Polski słownik biograficzny* [Polish biographical dictionary], vol. 14, p. 11. Krakow: Polish Academy of Sciences, 1969.

"Double Helix Serves Double Duty." *New York Times*, January 29, 2013, Science section, p. D3.

Durett, R., and D. Schmidt. "Waiting for Regulatory Sequences to Appear." *Annals of Applied Probability* 17 (2007): 1–32.

———. "Waiting for Two Mutations: With Applications to Regulatory Sequence Evolution and the Limits of Darwinian Evolution." *Genetics* 180 (2008): 1501–1509.

Dyson, Freeman. *Disturbing the Universe*. New York: Harper & Row, 1979.

"Earth's Moon a Rare Species." *Cosmos*, November 23, 2007. http://www.cosmosmagazine.com/news/earths-moon-a-rare-species.

Emelyanenko, V. V., D. J. Asher, and M. E. Bailey. "The Fundamental Role of the Oort Cloud in Determining the Flux of Comets through the Planetary System." *Monthly Notices of the Royal Astronomical Society* 381, no. 2 (2007): 779–89.

The ENCODE Project Consortium. "An Integrated Encyclopedia of DNA Elements in the Human Genome." *Nature* 489, no. 7414 (2012): 57–74.

Erwin, Douglas, Marc Laflamme, Sarah Tweedt, Erik Sperling, Davide Pisani, and Kevin Peterson. "The Cambrian Conundrum: Early Divergence and Later Ecological Success in the Early History of Animals." *Science* 334 (November 25, 2011).

Expelled: No Intelligence Allowed, directed by Nathan Frankowski. DVD. Vivendi Entertainment, 2008.

"Extraterrestrial Nucleobases in the Murchison Meteorite." *Earth and Planetary Science Letters* 270 (2008): 130–36. http://astrobiology.gsfc.nasa.gov/analytical/PDF/Martinsetal2008.pdf.

Feduccia, Alan. *The Origin and Evolution of Birds*, 2nd ed. New Haven, CT: Yale University Press, 1999.

Feynman, Richard. Cornell Lectures. http://amiquote.tumblr.com/post/4463599197/richard-feynman-on-how-we-would-look-for-a-new-law.

"52 Years and $750 Million Prove Einstein Was Right." *New York Times*, May 4, 2011.

Flight: The Genius of Birds, directed by Lad Allen. DVD. Illustra Media, 2013.

Francis, Matthew. "Quantum Entanglement Shows That Reality Can't Be Local." *Ars Technica*, October 30, 2012.

Flannery, Michael A. *Alfred Russel Wallace: A Rediscovered Life*. Seattle: Discovery Institute Press, 2011.

Frazzetta, Thomas H. *Complex Adaptations in Evolving Populations*. Stamford, CT: Sinauer Associates, 1975.

Freeland, Stephen J., Robin D. Knight, Laura F. Landweber, and Laurence D. Hurst. "Early Fixation of an Optimal Genetic Code." *Molecular Biology and Evolution* 17, no. 4 (2000): 511–18. http://mbe.oxfordjournals.org/content/17/4/511.full.

Gauger, A. K., and D. D. Axe. "The Evolutionary Accessibility of New Enzyme Functions: A Case Study from the Biotin Pathway." *BIO-Complexity* 2, no. 1 (2011): 1–17.

Gauger, Ann, Douglas Axe, and Casey Luskin. *Science and Human Origins*. Seattle: Discovery Institute Press, 2012.

Gibbs, W. Wayt. "The Unseen Genome: Gems among the Junk." *Scientific American*, November 2003. http://www.detectingdesign.com/.../The%20Unseen%20Genome.doc.

Gingerich, Owen. *The Book Nobody Read: Chasing the Revolutions of Nicholaus Copernicus*. New York: Walker, 2004.

Gomes, R., H. F. Levison, K. Tsiganis, and A. Mrbidelli. "Origin of the Cataclysmic Late Heavy Bombardment Period of the Terrestrial Planets." *Nature* 435, no. 7041 (2005): 466–69.

Gonzalez, Guillermo, and Jay Richards. *The Privileged Planet: How Our Place in the Cosmos Is Designed for Discovery*. Washington, DC: Regnery, 2004.

Gould, Stephen Jay. "Is a New and General Theory of Evolution Emerging?" *Paleobiology* 6, no. 1 (January 1980): 127.

———. "The Return of Hopeful Monsters." *Natural History* 86, no. 6 (June–July 1977): 24.

Grant, Peter R., and B. Rosemary Grant. "Predicting Microevolutionary Responses to Directional Selection on Heritable Variation." *Evolution* 49 (1995): 241–51.

———. "Speciation and Hybridization of Birds on Islands." In *Evolution on Islands*, ed. Peter R. Grant. Oxford: Oxford University Press, 1998.

Hart, Michael H. *The 100: A Ranking of the Most Influential Persons in History.* New York: Carol Publishing Group/Citadel Press, 1978; repr., 1992.

Hawking, Stephen Hawking. *A Brief History of Time.* New York: Bantam Books, 1988.

———. Interview with Ken Campbell. *Reality on the Rocks: Beyond Our Ken.* http://en.wikiquote.org/wiki/Stephen_Hawking.

Hecht, Jeff. "Solar Systems Like Ours May Be Rare." *New Scientist*, August 7, 2008. http://www.newscientist.com/article/dn14492-solar-systems-like-ours-may-be-rare.html.

Henry, Richard Conn. Review of *Quantum Enigma*, by Richard Rosenblum and Fred Kuttner. *Journal of Scientific Exploration* 21, no. 185 (2006).

Herrmann, S., A. Senger, K. Möhle, M. Nagel, E. V. Kovalchuk, and A. Peters. "Rotating Optical Cavity Experiment Testing Lorentz Invariance at the 10^{-17} Level." *Physical Review D* 80, no. 100 (2009): 105011.

Hersh, Megan N., Rebecca G. Ponder, P. J. Hastings, and Susan M. Rosenberg. "Adaptive Mutation and Amplification in *Escherichia coli*: Two Pathways of Genome Adaptation under Stress." *Research in Microbiology* 155 (2004): 352–59.

Horner, J., B. W. Jones, and J. Chambers. "Jupiter Friend or Foe? III. The Oort Cloud Comets." *International Journal of Astrobiology* 9, no. 1 (2010): 1–10.

Hoyle, Fred. *The Intelligent Universe: A New View of Creation and Evolution.* New York: Holt, Rinehart & Winston, 1983.

———. *The Mathematics of Evolution.* Acorn Enterprises LLC, 1999; originally published in print form as *The Mathematics of Evolution*, Weston Publication on the Cosmic Origin of Life No. 1, 1987.

Huberman, Jack, ed. *The Quotable Atheist.* New York: Nation Books, 2007.

Iglesias-Groth, S., A. Manchado, R. Rebolo, J. I. Gonzalez Hernandez, D. A. Garcia-Hernandez, and D. L. Lambert. "A Search for Interstellar Anthracene toward the Perseus Anomalous Microwave Emission Region." *Monthly Notices of the Royal Astronomical Society* 407, no. 4 (May 2010): 2157.

James, Frances C., and John A. Pourtless IV. "Cladistics and the Origins of Birds: A Review and Two New Analyses." *Ornithological Monographs* 66 (2009): 1–78.

Jastrow, Robert. *God and the Astronomers*, 2nd ed. New York: Reader's Library, 2000.

Jimenez-Torres, Juan J., Barbara Pichardo, George Lake, and Henry Throop. "Effect of Different Stellar Galactic Environments on Planetary Discs: I. The Solar Neighbourhood and the Birth Cloud of the Sun." *Monthly Notices of the Royal Astronomical Society* 418 (2013): 1272–84.

Johnson, Phillip. "The Church of Darwin." *Wall Street Journal*, August 16, 1999.

Jurica, M. S., and Moore, M. J. "Pre-mRNA Splicing: Awash in a Sea of Proteins." *Molecular Cell* 12 (2003): 5–14.

Kant, Immanuel. *Critique of Pure Reason.* 1788. Edited and translated by P. Gruyer and A. W. Wood. Cambridge: Cambridge University Press, 1998.

Keller, Timothy. *The Reason for God: Belief in an Age of Skepticism.* New York: Riverhead Books, 2008.

Kelley, Deborah. "Is It Time to Throw Out 'Primordial Soup' Theory?" NPR, February 7, 2010.

Kelly, James. "If Only Charles Darwin Could See His Descendant Now." *National Catholic Register*, August 14, 2013. http://www.ncregister.com/daily-news/if-only-charles-darwin-could-see-his-descendant-now?utm_source=feedburner&utm_medium=feed&utm_campaign=Feed%3A+NCRegisterDailyBlog+National+Catholic+Register# When:2013-08-14%2014:32:01. Accessed September 2, 2013.

Khalturin, K., G. Hemmrich, S. Fraune, R. Augistin, and T. C. Bosch. "More Than Just Orphans: Are Taxonomically-Restricted Genes Important in Evolution?" *Trends in Genetics* 25, no. 9 (2009): 404–13.

Kitzmiller v. Dover Area School District, 400 F. Supp. 2d 707 (2005).

Knapp, Alex. "Physicists Quantum Teleport Photons Over 88 Miles." *Forbes*, September 6, 2012.

Koonin, Eugene V. *The Logic of Chance: The Nature and Origin of Biological Evolution*. Upper Saddle River, NJ: FT Press, 2011.

Koonin, E. V., and A. S. Novozhilov. "Origin and Evolution of the Genetic Code: The Universal Enigma." *IUBMB Life* 61, no. 2 (2009): 99–111. Abstract available at http://www.ncbi.nlm.nih.gov/pubmed/19117371. Accessed August 31, 2013.

Kopparapu, R. K., et al., "Habitable Zones around Main-Sequence Stars: New Estimates." *Astrophysics Journal* 765, no. 2 (2013): 131.

Krasinsky, G. A., E. V. Pitjeva, M. V. Vasilyev, and E. I. Yagudina. "Hidden Mass in the Asteroid Belt." *Icarus* 158, no. 1 (July 2002): 98–105.

Kuhn, Thomas S. *The Structure of Scientific Revolutions*. Chicago: University of Chicago Press, 1962.

Kushnir, Anna. "The Origins of Life on Earth. Really." March 9, 2009. http://blogs.nature.com/boston/2009/03/09/the-origins-of-life-on-earth-really.

Lewis, C. S. *Mere Christianity*. London: Macmillan, 1943.

———. *The Problem of Pain*. London: Macmillan, 1961.

Lewis, J. S. *Worlds Without End: The Exploration of Planets Known and Unknown*. Reading, UK: Helix Books, 1998.

Lineweaver, Charles, Yeshe Fenner, and Brad K. Gibson. "The Galactic Habitable Zone and the Age Distribution of Complex Life in the Milky Way." *Science* 303, no. 5654 (2004): 59–62.

Lockwood, G. W., B. A. Skiff, and R. R. Radick. "The Photometric Variability of Sun-like Stars: Observations and Results, 1984–1995." *Astrophysical Journal* 485 (1997): 789–811.

Long, Jeffrey, MD, with Paul Perry. *Evidence of the Afterlife: The Science of Near-Death Experiences.* New York: HarperOne 2010.

Luskin, Casey. "How Sudden Was the Cambrian Explosion?" *Evolution News and Views*, July 16, 2013. http://www.evolutionnews.org/2013/07/how_sudden_was_074511.html. Accessed July 17, 2013.

———. "Junk DNA and Science-Stopping." *Evolution News*. http://www.evolutionnews.org/2006/12/junk_dna_and_sciencestopping002886.html.

———. "Not Making the Grade: An Evaluation of 22 Recent Biology Textbooks and Their Use of Selection Icons of Evolution." September 26, 2011. http://www.evolutionnews.org/DiscoveryInstitute_2011TextbookReview.pdf.

Martin, Rebecca G., and Mario Livio. "On the Formation and Evolution of Asteroid Belts and Their Potential Significance for Life." *Monthly Notices of the Royal Astronomical Society* 428 (2013): L11–L15.

Martin, Robert A. *Missing Links: Evolutionary Concepts & Transitions Through Time.* Sudbury, MA: Jones and Bartlett, 2004.

Matthews, Robert. "Patients Near Death See Visions of Hell." *Daily Telegraph.* http://www.theforbiddenknowledge.com/hardtruth/visions_of_hell.htm.

McLatchie, Jonathan. "The Finely Tuned Genetic Code." *Evolution News and Views*, November 19, 2011. http://www.evolutionnews.org/2011/11/the_finely_tuned_genetic_code052611.html.

Mebane, Alexander. *Darwin's Creation-Myth.* Venice, FL: P&D Printing, 1994.

Melott, A. L., and B. C. Thomas. "Astrophysical Ionizing Radiation and Earth: A Brief Review and Census of Intermittent Sources." *Astrobiology* 11 (2011): 343–61.

"Memories of Near Death Experiences: More Real Than Reality?" *Science Daily*, March 27, 2013. http://www.sciencedaily.com/releases/2013/03/130327190359.htm.

Meyer, Stephen. *Darwin's Doubt.* New York: HarperOne, 2013.

————. *Signature in the Cell: DNA and the Evidence for Intelligent Design.* New York: HarperOne, 2009.

Miller, Stanley L., and H. James Cleaves, "Prebiotic Chemistry on the Primitive Earth." In *Systems Biology: Volume 1. Genomics*, edited by Isidore and Gregory Stephanopoulos. New York: Oxford University Press, 2007.

Morowitz, Harold. *Energy Flow in Biology.* New York: Academic Press, 1968.

Moskowitz, Clara. "Solar Systems Like Ours May Be Rare." Space.com, July 21, 2008. http://www.space.com/5638-solar-systems-rare.html.

Munn, Colin. *Marine Microbiology: Ecology and Applications*, 2nd ed. New York: Garland Science, 2011.

Nagel, Thomas. "A Philosopher Defends Religion." *New York Times Review of Books*, September 27, 2012.

NASA Science. "Ten-Thousandth Near-Earth Object Discovered." http://science.nasa.gov/science-news/science-at-nasa/2013/24jun_neo/. Accessed June 25, 2013.

Neal, Mary C., MD. *To Heaven and Back.* Colorado Springs, CO: WaterBrook Press, 2011.

Orgel, Leslie E. *The Origins of Life: Molecules and Natural Selection.* London: Chapman & Hall, 1973.

Palmer, Jason. "Kepler Telescope: Earth-Sized Planets 'Number 17bn.'" BBC News, January 8, 2013. http://www.bbc.co.uk/news/science-environment-20942440.

Panek, Richard. *The 4 Percent Universe: Dark Matter, Dark Energy, and the Race to Discover the Rest of Reality.* New York: Houghton Mifflin, 2011.

Parker, Joe, Georgia Tsagkogeorga, James A. Cotton, Yuan Liu, Paolo Provero, Elia Stupka, and Stephen J. Rossiter. "Genome-Wide Signatures of Convergent Evolution in Echolocating Mammals." *Nature* (2013). http://www.nature.com/nature/journal/vaop/ncurrent/full/nature12511.html?WT.ec_id=NATURE-20130905. Accessed September 6, 2013.

Pearson, Helen. "Genetic Information Codes and Enigmas." *Nature* 444, no. 259 (November 16, 2006).

Penrose, Roger. *The Road to Reality: A Complete Guide to the Laws of the Universe.* London: Random House, 2004.

Pfalzner, S. 2013, "Early Evolution of the Birth Cluster of the Solar System." *Astronomy & Astrophysics* 549 (January 2013): A82

Philip, G. K., and S. J. Freeland. "Did Evolution Select a Nonrandom 'Alphabet' of Amino Acids?" *Astrobiology* 11, no. 3 (April 2011): 235–40.

Planck, Max. *Scientific Autobiography and Other Papers*, trans. F. Gaynor. New York: Philosophical Library, 1949.

Plantinga, Alvin. *Where the Conflict Really Lies: Science, Religion, and Naturalism.* Oxford: Oxford University Press, 2012.

Podos, Jeffrey, and Stephen Nowicki. "Beaks, Adaptation, and Vocal Evolution in Darwin's Finches." *BioScience* 54, no. 6 (June 2004): 501–10.

"Pretty Useful: Appendix Evolved More Than 30 Times." *Science Now*, February 13, 2013. http://www.wired.com/wiredscience/2013/02/appendix-revolution/?cid=co5811204.

"Quantum Wonders: Corpuscles and Buckyballs." *New Scientist*, May 6, 2010.

Quick, Devon E., and John A. Ruben. "Cardio-Pulmonary Anatomy in Theropod Dinosaurs: Implications from Extant Archosaurs." *Journal of Morphology* 270 (2009): 1232–46.

"Rare Japanese Plant Has Largest Genome Known to Science." *Science Daily*, October 7, 2010. http://www.sciencedaily.com/releases/2010/10/101007120641.htm.

"Reality Check." *New Scientist*, August 3–9, 2013, p. 34.

Redfern, Simon. "Earth Life 'May Have Come from Mars.'" BBC News. http://www.bbc.co.uk/news/science-environment-23872765. Accessed August 30, 2013.

Rees, Martin. *Just Six Numbers: The Deep Forces That Shape the Universe.* London: Weidenfeld & Nicolson, 1999.

Reidhaar-Olson, John, and Robert Sauer. "Functionally Acceptable Substitutions in Two Alpha-Helical Regions of Lambda Repressor." *Proteins: Structure, Function, and Genetics* 7, no. 4 (1990): 306–16.

"Relative Differences: The Myth of 1%." *Science* 316 (June 29, 2007): 1836.

Reynolds, John Mark. "Thoughts on Faith and Science (Part I)." http://www.patheos.com/blogs/eidos/2013/07/thoughts-on-faith-and-science-part-i/. Accessed July 17, 2013.

"Right-Handed Amino Acids Help Bacteria Adapt." Howard Hughes Medical Institute. http://www.hhmi.org/news/right-handed-amino-acids-help-bacteria-adapt. Accessed August 28, 2013.

Romjue, Nickell John. *I, Charles Darwin*. Tucson, AZ: Whatmark, 2011.

"Rover Finds Hint of Life's Cradle on Mars." *New Scientist*, March 23–29, 2013.

Rowan-Robinson, Michael. *Cosmology*, 3rd ed. Oxford: Clarendon Press, 1996.

Royal Society. "Newton Beats Einstein in Polls of Royal Society Scientists and the Public." http://royalsociety.org/News.aspx?id=1324&terms=Newton+beats+Einstein+in+polls+of+scientists+and+the+public.

Sato, Naoki. "Comparative Analysis of the Genomes of Cyanobacteria and Plants." *Genome Informatics* 13 (2002): 173–82.

Schopf, J. W. "Microfossils of the Early Archean Apex Chert." *Science* 260 (April 30, 1993).

Schopf, J. W., and B. M. Packer. "Early Archean (3.5-Billion- to 3.3-Billion-Year-Old) Microfossils from Warrawoona Group, Australia." *Science* 237, no. 4810 (1987): 70.

Schram, David. "Dark Matter and the Origin of Cosmic Structure." *Sky & Telescope* (October 1994): 29.

Schutzenberger, Marcel P. "Mathematical Challenges to the Neo-Darwinian Interpretation of Evolution." In *Proceedings of the Wistar Symposium*. Philadelphia: Wistar Institute Press, 1967.

Schwartz, Jeffrey H. *Sudden Origins: Fossils, Genes, and the Emergence of Species*. New York: Wiley, 1999.

Shapiro, James. "How Life Changes Itself: The Read-Write (RW) Genome." *Physics of Life Reviews* 10, no. 3 (2013): 287–323.

Shapiro, Robert. *Origins: A Skeptic's Guide to the Creation of Life on Earth*. New York: Summit Books, 1996.

Simmons, Geoffrey, MD. *Billions of Missing Links: A Rational Look at the Mysteries Evolution Can't Explain*. Eugene, OR: Harvest Hour Publishers, 2007.

Simola, Daniel F., et al. "Social Insect Genomes Exhibit Dramatic Evolution in Gene Composition and Regulation While Preserving Regulatory Features Linked to Sociality." *Genome Research* 23, no. 8 (2013): 1235–47.

Slack, A., P. C. Thornton, D. B. Magner, S. M. Rosenberg, and P. J. Hastings. "On the Mechanism of Gene Amplification Induced under Stress in *Escherichia coli*." *PLoS Genetics* 2, no. 4 (2006): 385–98.

Stanley, Steven M. *Macroevolution: Pattern and Process*. New York: W. H. Freeman, 1979.

Steiner, Mark. *The Applicability of Mathematics as a Philosophical Problem*. Cambridge, MA: Harvard University Press, 1998.

Strahler, Arthur. *Science and Earth History: The Evolution/Creation Controversy*. Amherst, NY: Prometheus Books, 1987.

Suarez, Antoine, and Peter Adams. *Is Science Compatible with Free Will? Exploring Free Will and Consciousness in the Light of Quantum Physics and Neuroscience*. New York: Springer, 2013.

Surn, Garret, et al. "The Genome Sequence of the Leaf-Cutter Ant *Atta cephalotex* Reveals Insights into Its Obligate Symbiotic Lifestyle." *PLoS Genetics* 7 (2011): e1002007.

Swisher, III, Carl C., Yuan-qing Wang, Xiao-lin Wang, Xing Xu, and Yuan Wang. "Cretaceous Age for the Feathered Dinosaurs of Lianoing, China." *Nature* 400 (July 1, 1999): 58–61.

Tedesco, E. F., and F.-X. Desert. "The Infrared Space Observatory Deep Asteroid Search." *Astronomical Journal* 123, no. 4 (2002): 2070–82.

Tegmark, Max, Anthony Aguirre, Martin Rees, and Frank Wilczek. "Dimensionless Constants, Cosmology, and Other Dark Matter." *Physical Review Part D* (January 2006).

Vollmers, Christopher, Robert J. Schmitz, Jason Nathanson, Gene Yeo, Joseph R. Ecker, and Satchidananda Panda. "Circadian Oscillations of Protein-Coding and Regulatory RNAs in a Highly Dynamic Mammalian Live Epigenome." *Cell Metabolism* 17, no. 6 (December 2012): 833–45.

Ward, Peter D., and Donald Brownlee. *Rare Earth: Why Complex Life Is Uncommon in the Universe*. New York: Copernicus Books, 2004.

Webb, Richard. "Primordial Broth of Life Was a Dry Martian Cup-a-Soup." *New Scientist*, August 29, 2013. http://www.newscientist.com/article/dn24120-primordial-broth-of-life-was-a-dry-martian-cupasoup.html#.UiBpZbjD-Ul. Accessed August 30, 2013.

Webb, Stephen. *If the Universe Is Teeming with Aliens . . . Where Is Everybody? Fifty Solutions to Fermi's Paradox and the Problem of Extraterrestrial Life*. New York: Copernicus Books, 2002.

Weinberg, Steven. *The First Three Minutes: A Modern View of the Origin of the Universe*. New York: Basic Books, 1979.

Weiner, Jonathan. *The Beak of the Finch*. New York: Vintage Books, 1994.

Whitehead, Alfred North. "Religion and Science." *Atlantic*, August 1925.

"Why Some Scientist Embrace the 'Multiverse': This Universe's Evidence Suggests a Designing Intelligence, so Atheists Resort to an Idea with No Evidence." *National Review Online*, June 18, 2013. http://www.nationalreview.com/article/351319/why-some-scientists-embrace-multiverse-dennis-prager. Accessed July 18, 2013.

"The World's Tallest Trees." *National Geographic*, December 2012.

Wright, N. T. *The Resurrection of the Son of God*. Vol. 3 of *Christian Origins and the Question of God*. Minneapolis, MN: Fortress, 2003.

Yong, E. "ENCODE: The Rough Guide to the Human Genome." *Discover Magazine*, September 5, 2012.

INDEX

new laws of, 85
Newton, 183–184
nuclear physicists, 67
Planck, Max, 43
Plank time, 52
Planets. *See also* specific planets
  Aristotle, shell model of, 33
  brightness of, 34
  carbon, 181
  carbon dioxide, 182
  Copernicus, theory of, 34
  formation, 181
  gravity, Galileo and, 35
  iron, molten, 181
  magnetic field, 181
  metal poor, 172
  observation of motion, 36n
  orbits, Kepler and, 35
  plate tectonics, 181–182
  solar wind, 181
  water, 181, 182
Plantinga, Alvin, 26
Plate tectonics, 181–182
Pluto, 175
Poe, Edgar Allen, 68
Polar cod freezing, 161
Pope Clement VII, 34
Post-World War II theories, 198–199
Predictions vs. facts, 144
Predictive value, of theories, 201
Primordial soup, 96
The Privileged Planet, 170, 181
Probability. *See also* Mathematics/numbers;
  Religion; scientism
  and religion, 53–55
  technology and, 136–138
Proof by contradiction, 221–222
Proofreading mechanisms, of life, 135–136
Proteins, structure of, 129. *See also* Amino
  acids; (deoxyribonucleic acid).
  as building blocks of life, 92, 94, 95, 106
  carboxyl group, 131
  chemical properties, different, 132

complexity of, 129, 131–132
functional, 92, 106–107, 153–154, 157–158
left-handed molecule, 132
length, 131
Miller-Urey experiment, 95
mirror-image shapes, 132
nanotechnology, 132
optical isomers, 132
side train, of atoms, 131–132
simple, 106
three-dimensional folding, 132
Proxima Centauri, 177
Ptolemaic system, 33, 36
Punctuated equilibrium, 149

Quantas, 186–187
Quantum field, 71
Quantum gravity, 191–192. *See also* Gravity
Quantum physics, 11, 186–193. *See also*
  Physics
  circle of existence, 193
  Eigen functions, 186
  entanglement experiment, 188–189
  Hamiltonian operators, 186
  Hilbert spaces, 186
  Klein-Gordon equations, 186
  Lie algebras, 186
  microscopic world, 186, 191–192
  non-material agency outside of space-time, 193
  observation, changes to quantum system, 189–190
  one-slit/two-slit experiment, 187–188
  outside of time/space connections, 189, 190, 193, 201
  phase state, 79–80, 188
  quantas, 186–187
  Schrödinger's wave equation, 186
  space-time, 192
  spooky action at a distance, 189
  string theory, 192
  uncertainty principle, 186
  universe as constructed illusion, 190

ACKNOWLEDGEMENTS

I am grateful to so many for making this book possible. Special thanks to Peter Fisher, who heads up the Department of Physics at Massachusetts Institute of Technology. We have had many conversations on the subjects of this book, and he gave me invaluable help. What I remember most was once when we were having dinner, I commented that someone should write a book explaining why science agreed with religion, and Peter suggested maybe that someone could be me. Carole Sargent encouraged me to follow my dream, what seems to have become my calling. Bill Dembski inspired me at a debate in 2005, and more recently he allowed me to join an online group of scientists and others seeking wonder. You will encounter many of the things I learned from that group in this book. I am grateful to the scientists and technical people who helped me, especially Jerry Bergman, Ann Gauger, Guillermo Gonzalez, and Casey Luskin, each of whom generously read through the bulk of the manuscript and made invaluable suggestions. Casey helped me in many other ways; he encouraged me to add my personal beliefs in the last chapter. Thanks also to Richard Willing, my son Matt, my brother Ted, and my mother, who also read parts of the manuscript and gave me suggestions. Thanks to Rob Sheldon, for his insightful contributions. Thanks to Owen Gingerich, for his helpful comments.

I am grateful to my publisher, John Groom, at Attitude Media, for his patience and his many talents to make this dream come true.

Thanks also to the many persons of faith who inspired me, from Michael Harmon at my first Episcopal Church, in Arlington, Virginia, over thirty years ago, to Bob Edmonds and John Schule on Martha's Vineyard, to Walter ("Frisby") Hendricks and Chris Rodriguez in Florida.

Finally, I am grateful for the Discovery Institute, a nonprofit public policy think tank. To me, the Discovery Institute is a candle of free thought and reason flickering in our overwhelmingly materialistic culture.

Made in the USA
Lexington, KY
23 November 2014